Columbia University

Contributions to Education

Teachers College Series

No. 80

AMS PRESS
NEW YORK

Measurements of Some Achievements in Arithmetic

By Clifford Woody, Ph.D.
Associate Professor of Education
University of Washington

TEACHERS COLLEGE, COLUMBIA UNIVERSITY
CONTRIBUTIONS TO EDUCATION, NO. 80

REVISED AND ENLARGED EDITION

Published by
Teachers College Columbia University
New York City
1920

Library of Congress Cataloging in Publication Data

Woody, Clifford, 1884-1948.
Measurements of some achievements in arithmetic.

Reprint of the 1920 ed., issued in series: Teachers College, Columbia University. Contributions to education, no. 80.
Originally presented as the author's thesis, Columbia.
Includes bibliographical references.
1. Arithmetic--Study and teaching. 2. Mathematical ability--Testing. I. Title. II. Series: Columbia University. Teachers College. Contributions to education, no. 80.
QA135.5.W57 1972 372.1'2'62 73-177622
ISBN 0-404-55080-0

Reprinted by Special Arrangement with Teachers College Press, New York, New York

From the edition of 1920, New York
First AMS edition published in 1972
Manufactured in the United States

AMS PRESS, INC.
NEW YORK, N. Y. 10003

PREFACE

The revision of the monograph, *Measurements of Some Achievements in Arithmetic,* is due to several causes: (1) The second impression of the original monograph itself is exhausted while the demand for it is still present. (2) The extensive use of the tests has caused an ever-increasing demand for the latest information concerning the standards obtained in the various places where the tests have been given, and it seemed best to make them available through this publication. (3) The author has become convinced from his work in educational measurements that more specific directions should be given concerning the administration of the tests and the interpretation and presentation of the results. Many people are able to give tests but have no idea how to treat the results so as to get the greatest amount of meaning from them. Thus it seemed advisable to revise the first part of the original monograph and to suggest methods of treating the results so as to make the scales more valuable to those using them.

The new edition contains the latest standards for the scales and practical suggestions for measuring the value of measurement with them. Such is the content of Part I. Part II, giving the history and derivation of the original scales, is a reproduction from the original monograph. It is possible for those who are interested only in how to make best use of the tests, to procure Part I as a separate bulletin; but since there is a wide demand for the detailed discussion of the derivation of the scales it seems advisable to make available to those desiring it an enlarged edition giving both the suggestions on how to use the scales and a history of the derivation.

C. W.

CONTENTS

PART I

THE USE OF THE SCALES

PART II

THE DERIVATION OF THE SCALES

PART I

THE USE OF THE SCALES

1. A Brief Description of the Scales and Their Development

The scales as presented in this section consist of two series, A and B, for each of the four fundamental processes in arithmetic. Series A and B are similar in construction, but Series B contains only about half the problems of Series A, and they are so selected that they come at approximately equal intervals on the scale. Series B, because of its greater brevity, is especially useful when the time for testing is limited; but Series A, because of the greater number and variety of problems contained, is more valuable from the standpoint of diagnosis.

These scales represent the final product gained from testing from 20,000 to 30,000 school children in Indiana, New Jersey, Connecticut, and New York. The preliminary tests consisted of four sheets of problems, one for each of the fundamental operations. In devising different sheets, there was a definite attempt to select as great a variety of problems as the fundamental operations would permit, and to select them with various degrees of difficulties ranging from the easiest combinations that could be found to problems so difficult that they could be solved by only a very small percentage of the pupils in the eighth grade. In the first preliminary set of test sheets, little attention was paid to the social utility of the different exercises, but as experience was gained, it became evident that each and every problem to be selected for the final scales must represent the type of problem frequently occurring in our complex social life. Consequently, after making a survey of the different investigations conducted to determine the actual needs for arithmetic in everyday transactions, i.e., the types of problems solved, the particular transactions used, etc., the original lists of problems were carefully scrutinized and all problems were rejected which did not represent common social needs. Whenever new problems were added, the social criterion was always taken into consideration. The task finally resolved itself into taking a series of problems representing a cross section of the minimal essentials for that operation, expressed in problems of as many types as possible, and arranging them in order of their degree of difficulty.

The final arrangement as given in the present scales was determined by giving the last revision of the preliminary tests to about 20,000 children and subjecting the results to intricate statistical treatment. Thus, the position of the different problems in the scales depends upon the percentage of the children correctly solving them referred to the normal curve of distribution.[1]

The simple fact that these problems are arranged in the form of scales makes these tests different from all other standardized tests in arithmetic, in that the measurement of ability is in terms of the "difficulty" rather than in terms of "how-many-ness." Such a characteristic has caused investigators to designate the scales by such names as "difficulty tests," "development tests," or "power tests."

2. Directions for Administering the Tests

These scales are useful as measures of achievement of a class, or of a whole school system. Series B is recommended where the amount of time for testing is limited, but Series A is better for diagnostic purposes. The same test is given to all children in the different grades. The addition and subtraction scales can be used in grades two to eight inclusive; the multiplication and division scales, in grades three to eight inclusive. The scales may be given in any order, in immediate succession, or with convenient intervals. It is recommended that all tests of Series B be given in succession, but two tests of Series A may well be given on one day, and the remaining two on the next.

If the measurements with these scales are to be valid and comparable, it is necessary that the same standardized procedure be followed in giving them as was followed in their original development. The main points to be observed are:

[1] For a more extended treatment, reference must be made to Part II of this monograph.

1. Clear the desks, provide each pupil with a sharpened pencil, and say, "One of these sheets will be laid on your desk face down. Leave it face down until I tell you to turn it."

2. Have the pupils in the front seats distribute the tests.[1] Then say, "Get your pencils. Turn the sheet and answer the question at the top." (The number of questions to be answered can be determined by the one giving the tests. It will take less time and cause less confusion if the one giving the tests will repeat the question and tell the children what to write. For example, say, "The first line asks, 'What is your name?' Write your name," etc.)

3. When all desired information has been secured, then say, without specially stressing any portion thereof, "Every problem on the sheet which I have given you is an addition problem, an 'and problem.' Work as many of these problems as you can and be sure that you get them right. Do all of your work on this sheet of paper and don't ask anybody any questions. Begin."

4. For each scale in Series A, allow *twenty minutes*; for each in Series B, allow *ten minutes*. See that all children quit work when the signal "STOP" is given. (Most of the children will have finished before that time. Those who have not, in all probability have done all they can; at least they have taken as much time as it takes the average class to complete the test.)

The only variation in procedure in giving any of the other tests is the substitution in the formula of specific directions of the expressions subtraction or "take away problems," multiplication or "times problems," and division or "into problems," for the expression addition or "and problems." The expressions "and," "take away," "times," and "into" problems are used so as to make clear to the children what process is to be involved. It is possible that teachers use these expressions in the lower grades instead of "addition, subtraction, multiplication and division problems." There is a great variation in the names applied to the subtraction process. When the original tests were given the teacher's terminology was used.

---

[1] Scales may be purchased from the Bureau of Publications, Teachers College, Columbia University.

SERIES A[1]

ADDITION SCALE

Name...

When is your next birthday?...............................How old will you be?...............

Are you a boy or a girl?.......................In what grade are you?...........................

(1)
2
3

(2)
2
4
3

(3)
17
2

(4)
53
45

(5)
72
26

(6)
60
37

(7)
3+1=

(8)
2+5+1=

(9
20
10
2
30
25

(10)
21
33
35

(11)
32
59
17

(12)
43
1
2
13

(13)
23
25
16

(14)
25+42=

(15)
100
33
45
201
46

(16)
9
24
12
15
19

(17)
199
194
295
156

(18)
2563
1387
4954
2065

(19)
$ .75
1.25
.49

(20)
$12.50
16.75
15.75

(21)
$8.00
5.75
2.33
4.16
.94
6.32

(22)
547
197
685
678
456
393
525
240
152

(23)
$\frac{1}{3}+\frac{1}{3}=$

(24)
4.0125
1.5907
4.10
8.673

(25)
$\frac{3}{8}+\frac{5}{8}+\frac{7}{8}+\frac{1}{8}=$

(26)
$12\frac{1}{2}$
$62\frac{1}{2}$
$12\frac{1}{2}$
$37\frac{1}{2}$

(27)
$\frac{1}{8}+\frac{1}{4}+\frac{1}{2}=$

(28)
$\frac{3}{4}+\frac{1}{4}=$

(29)
$4\frac{3}{4}$
$2\frac{1}{4}$
$5\frac{1}{4}$

(30)
$2\frac{1}{2}$
$6\frac{3}{8}$
$3\frac{3}{4}$

(31)
113.46
49.6097
19.9
9.87
.0086
18.253
6.04

(32)
$\frac{3}{4}+\frac{1}{2}+\frac{1}{4}=$

(33)
.49
.28
.63
.95
1.69
.22
.33
.36
1.01
.56
.88
.75
.56
1.10
.18
.56

(34)
$\frac{1}{6}+\frac{3}{8}=$

(35)
2 ft. 6 in.
3 ft. 5 in.
4 ft. 9 in.

(36)
2 yr. 5 mo.
3 yr. 6 mo.
4 yr. 9 mo.
5 yr. 2 mo.
6 yr. 7 mo.

(37)
$16\frac{1}{3}$
$12\frac{1}{8}$
$21\frac{1}{2}$
$32\frac{3}{4}$

(38)
25.091+100.4+25+98.28+19.3614=

[1] The scales are printed in large type, on separate sheets, $8\frac{1}{2}''$ x $11''$, with ample space for the insertion of answers.

SERIES A

SUBTRACTION SCALE

Name...

When is your next birthday?..............................How old will you be?...............

Are you a boy or a girl?......................In what grade are you?..........................

(1)	(2)	(3)	(4)	(5)	(6)	(7)	(8)	(9)	(10)	(11)
8	6	2	9	4	11	13	59	78	7 — 4 =	76
5	0	1	3	4	7	8	12	37		60

(12)	(13)	(14)	(15)	(16)	(17)	(18)	(19)	(20)
27	16	50	21	270	393	1000	567482	$2\frac{3}{4}$ — 1 =
3	9	25	9	190	178	537	106493	

(21)	(22)	(23)	(24)	(25)	(26)
10.00	$3\frac{1}{2} - \frac{1}{2} =$	80836465	$8\frac{7}{8}$	27	4 yds. 1 ft. 6 in.
3.49		49178036	$5\frac{3}{4}$	$12\frac{5}{8}$	2 yds. 2 ft. 3 in.

(27)	(28)	(29)	(30)
5 yds. 1 ft. 4 in.	10 — 6.25 =	$75\frac{3}{4}$	9.8063 — 9.019 =
2 yds. 2 ft. 8 in.		$52\frac{1}{4}$	

(31)	(32)	(33)	(34)	(35)
7.3 — 3.00081 =	1912 6 mo. 8 da.	$\frac{5}{12} - \frac{2}{10} =$	$6\frac{1}{8}$	$3\frac{7}{8} - 1\frac{5}{8} =$
	1910 7 mo. 15 da.		$2\frac{7}{8}$	

Series A

MULTIPLICATION SCALE

Name..

When is your next birthday?..............................How old will you be?...............

Are you a boy or a girl?..........................In what grade are you?.......................

(1)	(2)	(3)	(4)	(5)	(6)	(7)
$3 \times 7 =$	$5 \times 1 =$	$2 \times 3 =$	$4 \times 8 =$	23	310	$7 \times 9 =$
				3	4	

(8)	(9)	(10)	(11)	(12)	(13)	(14)	(15)
50	254	623	1036	5096	8754	165	235
3	6	7	8	6	8	40	23

(16)	(17)	(18)	(19)	(20)	(21)	(22)
7898	145	24	9.6	287	24	$8 \times 5\frac{3}{4} =$
9	206	234	4	.05	$2\frac{1}{2}$	

(23)	(24)	(25)	(26)	(27)	(28)	(29)
$1\frac{1}{4} \times 8 =$	16	$\frac{7}{8} \times \frac{3}{4} =$	9742	6.25	.0123	$\frac{1}{8} \times 2 =$
	$2\frac{5}{8}$		59	3.2	9.8	

(30)	(31)	(32)	(33)	(34)
2.49	$\frac{12}{25} \times \frac{15}{32} =$	6 dollars 49 cents	$2\frac{1}{2} \times 3\frac{1}{2} =$	$\frac{1}{2} \times \frac{1}{2} =$
36		8		

(35)	(36)	(37)	(38)	(39)
$987\frac{3}{4}$	3 ft. 5 in.	$2\frac{1}{4} \times 4\frac{1}{2} \times 1\frac{1}{2} =$	$.0963\frac{1}{8}$	8 ft. $9\frac{1}{2}$ in.
25	5		.084	9

Series A

DIVISION SCALE

Name..

When is your next birthday?.................................How old will you be?................

Are you a boy or a girl?........................In what grade are you?............................

(1)	(2)	(3)	(4)	(5)	(6)
$3\overline{)6}$	$9\overline{)27}$	$4\overline{)28}$	$1\overline{)5}$	$9\overline{)36}$	$3\overline{)39}$

(7)	(8)	(9)	(10)	(11)	(12)
$4 \div 2 =$	$9\overline{)0}$	$1\overline{)1}$	$6 \times$ .... $= 30$	$2\overline{)13}$	$2 \div 2 =$

(13)	(14)	(15)	(16)	(17)
$4\overline{)24 \text{ lbs. } 8 \text{ oz.}}$	$8\overline{)5856}$	$\frac{1}{4}$ of $128 =$	$68\overline{)2108}$	$50 \div 7 =$

(18)	(19)	(20)	(21)	(22)
$13\overline{)65065}$	$248 \div 7 =$	$2.1\overline{)25.2}$	$25\overline{)9750}$	$2\overline{)13.50}$

(23)	(24)	(25)	(26)
$23\overline{)469}$	$75\overline{)2250300}$	$2400\overline{)504000}$	$12\overline{)2.76}$

(27)	(28)	(29)	(30)
$\frac{7}{8}$ of $624 =$	$.003\overline{).0936}$	$3\frac{1}{2} \div 9 =$	$\frac{3}{4} \div 5 =$

(31)	(32)	(33)
$\dfrac{5}{4} \div \dfrac{3}{5} =$	$9\frac{5}{8} \div 3\frac{3}{4} =$	$52\overline{)3756}$

(34)	(35)	(36)
$62.50 \div 1\frac{1}{4} =$	$531\overline{)37722}$	$9\overline{)69 \text{ lbs. } 9 \text{ oz.}}$

Series B

ADDITION SCALE

Name.. ....

When is your next birthday?..................................How old will you be?................

Are you a boy or a girl?.........................In what grade are you?...........................

(1)
2
3
—

(2)
2
4
3
—

(3)
17
2
—

(5)
72
26
—

(7)
3 + 1 =

(10)
21
33
35
—

(13)
23
25
16
—

(14)
25 + 42 =

(16)
9
24
12
15
19
—

(19)
$ .75
1.25
.49
———

(20)
$12.50
16.75
15.75
———

(21)
$8.00
5.75
2.33
4.16
.94
6.32
———

(22)
547
197
685
678
456
393
525
240
152
—

(23)
$\frac{1}{3} + \frac{1}{3} =$

(24)
4.0125
1.5907
4.10
8.673
———

(30)
$2\frac{1}{2}$
$6\frac{3}{8}$
$3\frac{3}{4}$
—

(33)
.49
.28
.63
.95
1.69
.22
.33
.36
1.01
.56
.88
.75
.56
1.10
.18
.56
———

(36)
2 yr. 5 mo.
3 yr. 6 mo.
4 yr. 9 mo.
5 yr. 2 mo.
6 yr. 7 mo.
———

(38)
25.091 + 100.4 + 25 + 98.28 + 19.3614 =

Series B

SUBTRACTION SCALE

Name..

When is your next birthday?..............................**How old will you be?**...............

Are you a boy or a girl?.......................**In what grade are you?**...........................

(1)	(3)	(6)	(7)
8 5 —	2 1 —	11 7 —	13 8 —

(9)	(13)	(14)	(17)
78 37 —	16 9 —	50 25 —	393 178 —

(19)	(20)	(24)	(25)
567482 106493 ——	$2\frac{3}{4} - 1 =$	$8\frac{7}{8}$ $5\frac{3}{4}$ —	27 $12\frac{5}{8}$ ——

(27)	(31)	(35)
5 yds. 1 ft. 4 in. 2 yds. 2 ft. 8 in. ——	$7.3 - 3.00081 =$	$3\frac{7}{8} - 1\frac{5}{8} =$

SERIES B

MULTIPLICATION SCALE

Name..

When is your next birthday?..............................How old will you be?...............

Are you a boy or a girl?.......................In what grade are you?..........................

(1)	(3)	(4)	(5)
$3 \times 7 =$	$2 \times 3 =$	$4 \times 8 =$	23 3 —
(8)	**(9)**	**(11)**	**(12)**
50 3 —	254 6 —	1036 8 —	5096 6 —
(13)	**(16)**	**(18)**	**(20)**
8754 8 —	7898 9 —	24 234 —	287 .05 —
(24)	**(26)**	**(27)**	**(29)**
16 $2\frac{5}{8}$ —	9742 59 —	6.25 3.2 —	$\frac{1}{8} \times 2 =$
(33)	**(35)**	**(37)**	**(38)**
$2\frac{1}{2} \times 3\frac{1}{2} =$	$987\frac{3}{4}$ 25 —	$2\frac{1}{4} \times 4\frac{1}{2} \times 1\frac{1}{2} =$	$.0963\frac{1}{8}$ .084 —

SERIES B

DIVISION SCALE

Name..

When is your next birthday?................................How old will you be?...............

Are you a boy or a girl?........................In what grade are you?...........................

(1) $3\overline{)6}$	(2) $9\overline{)27}$	(7) $4 \div 2 =$	(8) $9\overline{)0}$
(11) $2\overline{)13}$	(14) $8\overline{)5856}$	(15) $\frac{1}{4}$ of $128 =$	(17) $50 \div 7 =$
(19) $248 \div 7 =$	(23) $23\overline{)469}$	(27) $\frac{7}{8}$ of $624 =$	(28) $.003\overline{).0936}$
(30) $\frac{3}{4} \div 5 =$	(34) $62.50 \div 1\frac{1}{4} =$	(36) $9\overline{)69 \text{ lbs. } 9 \text{ oz.}}$	

3. Directions for Scoring the Tests

In scoring the tests the standard for marking a problem correct is absolute accuracy, and, wherever possible, reduction to its lowest terms. If the results are to be comparable with the results and values established in these scales, only those answers should be accepted as correct which are found in Table I. These are the answers which were accepted in the original development of the scales.

A few incorrect answers are also listed in order to offer less chance for variation in the scoring of the results.

TABLE I: ANSWERS TO PROBLEMS

PROBLEM	ADDITION	SUBTRACTION	MULTIPLICATION	DIVISION
1	5	3	21	2
2	9	6	5	3
3	19	1	6	7
4	98	6	32	5
5	98	0	69	4
6	97	4	1,240	13
7	4	5	63	2
8	8	47	150	0
9	87	41	1,524	1
10	89	3	4,361	5
11	108	16	8,288	6-1/2 not 6+1
12	59	24	30,576	1
13	64	7	70,032	6 lbs. 2 oz. not 6 2
14	67	25	6,600	732
15	425	12	5,405	32
16	79	80	71,082	31
17	844	215	29,870	7-1/7 not 7+1
18	10,969	463	5,616	5,005
19	$2.49	460,989	38.4	35-3/7 not 35+3
20	$45.00	1-3/4	14.35	12
21	$27.50	6.51	60	390
22	3,873	3	46	6.75
23	2/3	31,658,429	10	20-9/23; 20.3, not 20+9
24	18.3762	3-1/8	42	30,004
25	2, not 16/8 nor 2/1	14-3/8	21/32	210
26	125, not 123-4/2 = 2	1 yd. 2 ft. 3 in. not 63 in.	574,778	.23
27	7/8	2 yds. 1 ft. 8 in. not 81 in.	20.000	546
28	1 not 4/4 nor 1/1	3-3/4 or 3.75	.12054	31.2
29	12-1/4 not 11-3/4 = 1-1/4	23-1/2 not 23-2/4 = 1/2	1/4 not 2/8	7/18
30	12-5/8 not 11-13/8 = 1-5/8	.7873	89.64	3/20 or .15
31	217.1413	4.29919	9/40	2-1/12
32	1-1/2 not 6/4 nor 1-2/4 = 1/2	1 yr. 10 mo. 23 da.	$51.92 or 51 dol. 92 cts.	2-17/30
33	10.55	13/60	8-3/4	72-3/13 or 72.23
34	13/24	3-1/4 not 3-2/8 = 1/4	1/4	50.
35	10 ft. 8 in. or 10-2/3 ft.	2-1/4 not 2-2/8 = 1/4	24693-3/4	71-7/177 or 71.04
36	22 yrs. 5 mo. or 22-5/12 yrs.		17 ft. 1 in.	7 lbs. 11-2/3 oz.; 7 lb.s 9 oz.
37	82-17/24		15-3/16	
38	268.1324		.0080902-1/2 or .00809025	
39			79 ft. 1-1/2 in.	

TABLE II: SAMPLE SCORE SHEET[1]

PUPILS' NAMES	NO. OF PROBLEM 1	2	3	4	5	6	7	8	9	10	11	12	13	14	15	16	17	18	19	20	21	22	23	ETC.
No. Getting Each Problem																								
% Getting Each Problem																								

4. DIRECTIONS FOR DETERMINING THE CLASS SCORE

For the determination of the class score, two different methods have been derived. The first method was derived especially for use with Series A, where there is no definite attempt to place the problems on a linear scale with equal steps between them. By this method, after the problems have been marked as right or wrong, enter the results on a score sheet similar to the one given in Table II. Thus a complete record of the particular problems solved by each child is obtained.

[1] Copies of this score sheet may be obtained from the Bureau of Publications, Teachers College.

To complete the class score, find the number of pupils in the class that solved each problem correctly. Divide the number by the total number in the class so as to get for each problem the per cent of the class that solved it correctly. Since, in the development of these scales, that problem which can be solved correctly by just 50 per cent of the class is taken as the best measure of the achievement of the class, select those five problems which come nearest to being solved by just 50 per cent of the class.[1] Table III gives the established value for each problem in the different processes. From Table V find the amount that must be added to or subtracted from the values given in Table III for each of these selected problems to find what difficulty a problem would need be in order that just 50 per cent of the class could solve it. Take the average of these five determinations and let it represent the class score. This means that a problem of that difficulty can be solved by just 50 per cent of the class in question.

To illustrate the determination of the class score, the five problems in addition which came nearest to being solved by just 50 per cent of the pupils in a certain third grade class of 61 pupils were problems Nos. 14, 17, 16, 18 and 15. These problems were thus solved correctly by 54, 54, 48, 48 and 68 per cent of the class, respectively. Table V tells how much to add to or subtract from the established value given in Tables III and IV for each of the problems in order to estimate the value of a problem that would be solved correctly by just 50 per cent of the class. Thus

% SOLVING PROBLEM	PROB. NO.	EST. VALUE OF PROB.	ADD. OR SUB.		VAL. OF PROB. SOLVED BY 50%
54	14	4.85	+.15	=	5.00
54	17	5.59	+.15	=	5.74
48	16	5.52	—.07	=	5.45
48	18	5.73	—.07	=	5.66
68	15	4.97	+.70	=	5.67
			Average	=	5.50

[1] The work of scoring may be greatly economized by omitting the scoring and entering on the score sheet of the problems which will not figure in the determination of the 50 per cent right point. Thus in an eighth grade class the first twenty or more problems can most certainly be neglected. A little experience will teach the scorer what problems he needs to score for a given class.

TABLE III

ESTABLISHED VALUE OF EACH PROBLEM IN EACH SCALE

SERIES A

NO. OF PROBLEM	ADDITION	SUBTRACTION	MULTIPLICATION	DIVISION
1	1.23	1.06	.87	1.57
2	1.40	1.48	1.05	2.08
3	2.50	1.50	1.11	2.18
4	2.61	1.50	1.58	2.31
5	2.83	1.70	2.38	2.40
6	3.21	1.75	2.62	2.46
7	3.26	2.18	2.68	2.56
8	3.35	2.51	2.71	3.05
9	3.63	2.57	3.78	3.16
10	3.78	2.65	3.79	3.20
11	3.92	2.88	4.09	3.49
12	4.18	2.90	4.26	3.59
13	4.19	2.96	4.71	3.96
14	4.85	3.64	4.72	4.06
15	4.97	3.70	4.73	4.60
16	5.52	4.35	5.05	4.67
17	5.59	4.41	5.20	4.98
18	5.73	4.42	5.24	5.16
19	5.75	5.18	5.38	5.26
20	6.10	5.52	5.63	5.31
21	6.44	5.70	5.72	5.36
22	6.79	5.75	5.83	5.48
23	7.11	5.76	5.83	5.56
24	7.43	5.91	5.89	5.58
25	7.47	6.77	6.29	5.78
26	7.61	7.07	6.30	5.91
27	7.62	7.21	6.58	6.04
28	7.67	7.38	6.85	6.43
29	7.71	7.41	6.97	6.76
30	7.71	7.41	7.00	6.83
31	7.97	7.49	7.07	6.87
32	8.04	7.52	7.07	6.88
33	8.18	7.69	7.29	7.22
34	8.22	7.72	7.50	7.24
35	8.58	7.84	7.65	8.17
36	8.67		7.66	8.23
37	8.67		8.02	
38	9.19		8.53	
39			8.61	

TABLE IV

Established Value of Each Problem in Each Scale

SERIES B

Addition		Subtraction		Multiplication		Division	
NO. OF PROBLEM	VALUE	NO. OF PROBLEM	VALUE	NO. OF PROBLEM	VALUE	NO. OF PROBLEM	VALUE
1	1.23	1	1.06	1	.87	1	1.57
2	1.40	3	1.50	3	1.11	2	2.08
3	2.50	6	1.75	4	1.58	7	2.56
5	2.83	7	2.18	5	2.38	8	3.05
7	3.26	9	2.57	8	2.71	11	3.49
10	3.78	13	2.96	9	3.78	14	4.06
13	4.19	14	3.64	11	4.09	15	4.60
14	4.85	17	4.41	12	4.26	17	4.98
16	5.52	19	5.18	13	4.71	19	5.26
19	5.75	20	5.52	16	5.05	23	5.57
20	6.10	24	5.91	18	5.24	27	6.04
21	6.44	25	6.77	20	5.63	28	6.43
22	6.79	27	7.21	24	5.89	30	6.83
23	7.11	31	7.49	26	6.30	34	7.24
24	7.43	35	7.84	27	6.58	36	8.23
30	7.71			29	6.97		
33	8.18			33	7.29		
36	8.67			35	7.65		
38	9.19			37	8.02		
				38	8.53		

The average of these 5 determinations (5.50) represents better than either single measurement the degree of difficulty that a problem must have in order that just 50 per cent of this class can solve it correctly. The class score for any other class can be computed in a similar manner.

The second method for the determination of the class score was derived especially for Series B[1] where there was a definite attempt to place the problems on a linear scale with equal steps between them. This method introduces a certain amount of error, but for all practical purposes it is a satisfactory measure. By this method the median number of problems solved correctly is taken as the measure of the achievement of any class.

[1] While this second method was primarily derived for Series B, current practice shows that most superintendents use it for determining the class score for Series A.

TABLE V

For Use in Estimating the Degree of Difficulty Required in a Problem so That Just 50 Per Cent of the Class can Solve it Correctly

	SUBTRACT		ADD
10%	1.90	50%	0.00
11	1.82	51	.03
12	1.74	52	.07
13	1.67	53	.11
14	1.60	54	15
15	1.54	55	.19
16	1.48	56	.22
17	1.42	57	.26
18	1.36	58	.30
19	1.30	59	.34
20	1.25	60	.38
21	1.20	61	.41
22	1.15	62	.45
23	1.10	63	.49
24	1.05	64	.53
25	1.00	65	.57
26	.95	66	.61
27	.91	67	.65
28	.86	68	.70
29	.82	69	.74
30	.78	70	.78
31	.74	71	.82
32	.70	72	.86
33	.65	73	.91
34	.61	74	.95
35	.57	75	1.00
36	.53	76	1.05
37	.49	77	1.10
38	.45	78	1.15
39	.41	79	1.20
40	.38	80	1.25
41	.34	81	1.30
42	.30	82	1.36
43	.26	83	1.42
44	.22	84	1.48
45	.19	85	1.54
46	.15	86	1.60
47	.11	87	1.67
48	.07	88	1.74
49	.03	89	1.82
		90	1.90

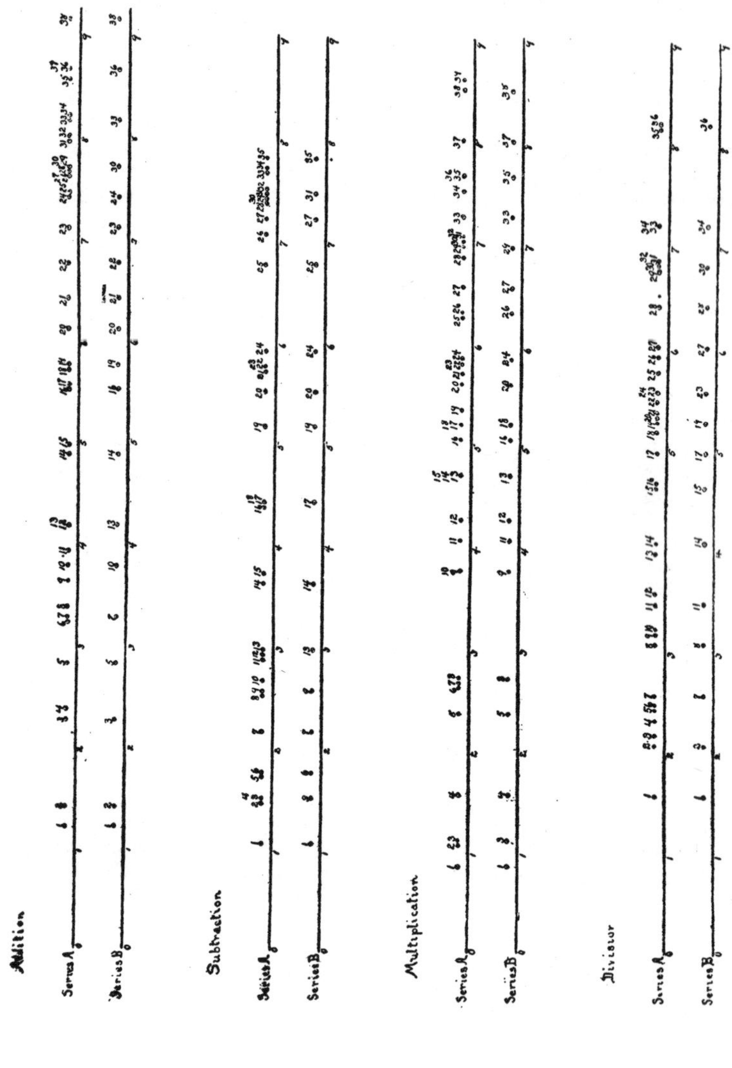

LINEAR PROJECTION OF THE DIFFICULTY OF THE PROBLEMS AS SHOWN BY TABLE III.

By the median number of problems solved is meant such a number of problems that there are just as many pupils who solve a greater number as there are pupils who solve a less number.

In order to determine the median point of the achievement of the class, it is necessary to make a distribution table, showing the number of pupils who are unable to solve a single problem correctly, the number who solved one problem, two problems, three problems, etc. The process of making the distribution table and computing the median score is exhibited in Table VI,

TABLE VI

SCORE SHEET OF A FIFTH GRADE CLASS (ADDITION, SERIES B)

SCORE	NUMBER OF PUPILS MAKING EACH SCORE		COMPUTATION OF MEDIAN
20			1. After checking the problems correctly solved, count the check marks in each paper, and indicate the number in the proper place in column 2.
19			2. Find the total number of scores (N).
18			3. Median equals middle score $= \frac{N}{2}$
17	II	2	Thus, $\frac{N}{2} = \frac{35}{2} = 17\frac{1}{2}$
16	IIII	4	Beginning at 1 in the third column and counting up, it is necessary to count $3\frac{1}{2}$ of the 7 to make $17\frac{1}{2}$, thus:
15	III	3	$17\frac{1}{2} = 1+1+2+3+4+3$ and $3\frac{1}{2}$ of the 7.
14	IIIII	5	Put $3\frac{1}{2}$ as a numerator over 7 and add to 13, the step on which the 7 occurs, thus:
13	IIIII II	7	MEDIAN $= 13 + \frac{3.5}{7} = 13.5$
12	III	3	
11	IIII	4	
10	III	3	
9	II	2	
8			
7	I	1	
6	I	1	
5			
4			
3			
2			
1			
0			
TOTAL SCORES (N)		35	
MEDIAN SCORE		13.5	

a score sheet for a fifth grade class. Since the process of computing the median as indicated on the score sheet is very brief and since many distributions are replete with irregularities, three additional distributions are presented in Table VII with a discussion of the method of determining the median point in each case.

TABLE VII

NUMBER OF ADDITION PROBLEMS CORRECTLY SOLVED BY THE MEMBERS OF THREE DIFFERENT CLASSES

	0	1	2	3	4	5	6	7	8	9	10	11	12	13	14	15	16	17	18
CLASS I ...	1	0	2	3	0	4	1	2	3	5	6	11	1	5	2	4	1	1	
CLASS II ...	3	3	4	4	7	5	3	4	3		1	1							
CLASS III ...		1	2	4	6	7	4	0	4	5	5	4	3	2	1				

According to these distributions 52 pupils are in Class I, 37 pupils in Class II, and 48 pupils in Class III. Let us proceed to find the median achievement for each of these class distributions.

Since there are 52 individuals in Class I the median point evidently falls between the achievements of the 26th and 27th pupils. Let us begin with the individual who was unable to solve a single problem correctly and count the two individuals who solved two problems, the three who solved three problems, and so on till we come to the step that includes the 26th individual. Now if we are to indicate the exact point in the achievement of the pupils where there are just as many pupils who solve a greater number of problems as there are those who solve a less number, it is necessary to count 5 of the 6 individuals who solved 10 problems correctly. Thus on the assumption that the individuals are distributed over any step at equal distances from one another, the median point is 5/6 of the distance through this step. Hence the median achievement of this class, i.e., the median number of problems solved, is 10.8 problems correctly solved.

Similarly, there are 37 pupils in Class II. The middle case is the 19th pupil, who is the fifth pupil in step 4. There are 18 pupils who solve a greater number of problems and 18 who solve a less number of problems. Thus the exact median point in the achievement of the class lies in the middle of that fraction of a step assigned to the 19th pupil. Thus the median point is $\frac{4.5}{7}$ of the distance through the 4th step. Hence the median achievement for this class is 4.6 problems solved correctly.

The distribution for Class III represents a peculiar difficulty in the calculation of its median. There are 48 pupils in this class and evidently the median point falls between the 24th and the 25th individual. However, it happens that 24 of the pupils solve more than seven problems and 24 of them solve less than seven. Probably the wisest assumption to make is that the 4 pupils on step 6 take up all of that step and the 4 pupils on step 8 take up all of that. If this is assumed, then the median falls on step 7, probably at 7.5 since any given distance on a scale is best represented by its middle point. Thus the median achievement for Class III is 7.5 problems solved correctly. By similar computations the medians of any distribution can be obtained. By the comparison of the medians thus determined, we get a very satisfactory measure of the achievement of any class on the basis of the total number of problems correctly solved.

5. Scores Attained in the Various Cities

In order that those giving the tests may have as much comparable data as possible, the scores attained in certain cities where the tests have been given are presented in Tables VIII, IX and X.

TABLE VIII

SUMMARY OF SCORES ATTAINED IN WOODY ARITHMETIC SCALES SERIES B

ADDITION

CITY	DATE[1]	SECTION	GRADES III	IV	V	VI	VII	VIII
Cherokee, Ia.	Feb. '17		10.7	13.5	15.6	16.6	16.7	16.5
Webster City, Ia.	Dec. '16		9.1	10.9	14.4	15.8	16.7	18.1
Hoquiam, Wash.	Mar. '17	B	8.0	12.6	14.3	15.2	16.8	17.4
		A	11.6	13.6	15.5	16.2	16.9	17.8
Seattle	June '18	B	....	....	14.6	15.7	16.6	17.2
		A	....	11.5	15.0	16.2	17.2	17.4
Kalama, Wn.	Nov. '17		7.8	11.5	13.5	14.3	15.4	....
Elyria, Ohio	Jan. '17		10.0	12.8	13.6	14.5	16.0	16.0
Pittsburgh, Pa.	Jan. '18		....	12.3	14.8	16.0	16.7	17.7
Trenton, N. J.	Apr. '18		7.9	10.3	12.1	13.9	....	....
Ellensburg, Wn.	Oct. '17		5.3	9.8	11.1	13.9	15.4	16.1
Duluth, Minn.	May '19	B	7.4	10.7	12.6	14.9	....	....
		A	9.5	12.0	14.1	15.7	....	....
Horace Mann	Nov. '16		9.0	12.9	14.5	15.8	16.1	16.5
(Columbia U.)	Apr. '17		12.4	14.1	12.1	16.9	16.6	17.5
N. Y. City School	Nov. '16		11.6	12.8	13.4	16.2	17.4	18.3
	Apr. '17		12.5	14.5	14.4	18.1	18.4	19.0
Denver, Colo.	'18	B	9.3	12.9	13.0	14.1	15.8	16.0
		A	9.8	13.2	13.0	14.5	15.6	15.7
Idaho Springs, Colo.	'18	B	....	....	12.9	12.9	14.7	15.5
	Average		9.5	12.3	13.7	15.3	15.8	17.0

SUBTRACTION

CITY	DATE	SECTION	III	IV	V	VI	VII	VIII
Cherokee, Ia.			7.2	9.6	11.2	13.6	13.8	13.8
Webster City, Ia.			6.8	9.0	11.1	12.7	13.9	14.4
Hoquiam, Wn.		B	8.3	9.3	11.2	13.4	14.3	15.0
		A	9.3	10.2	11.8	13.8	14.7	15.0
Seattle		B	....	....	11.0	12.0	13.3	13.9
		A	....	9.3	11.6	13.0	13.9	14.2
Kalama, Wn.			5.8	9.4	11.4	12.3	13.3	....
Elyria, Ohio			6.5	9.0	10.6	12.7	13.0	14.0
Pittsburgh, Pa.			....	9.4	12.0	13.2	14.1	14.9
Trenton, N. J.			6.1	8.3	9.6	11.8	....	....
Ellensburg, Wn.			4.9	7.5	8.8	10.8	12.5	13.4
Duluth, Minn.		B	6.0	8.6	10.5	12.6	....	....
		A	6.0	9.4	11.5	13.4	....	....
Horace Mann	Nov. '16		8.0	9.6	10.5	12.4	12.8	12.6
(Columbia U.)	Apr. '17		9.5	10.6	9.3	13.0	13.7	14.2
N. Y. City School	Nov. '16		8.5	9.5	11.2	13.0	14.7	14.7
	Apr. '17		9.1	10.7	12.1	14.2	15.0	15.0
Denver, Colo.		B	8.1	9.1	10.9	12.0	15.0	13.8
		A	7.0	10.0	10.6	11.0	13.5	14.5
Idaho Springs, Colo.		B	....	....	9.9	10.9	13.2	13.5
	Average		7.3	9.3	10.7	12.6	13.7	14.2

[1]The time of giving the tests will not be repeated save in the case of the Horace Mann and the New York City School where the tests were given twice in the same school year.

TABLE VIII—CONTINUED

MULTIPLICATION

CITY	DATE	SECTION	III	IV	V	VI	VII	VIII
Cherokee, Ia.			4.4	10.3	13.2	15.8	16.8	16.8
Webster City, Ia.			7.0	9.4	11.9	14.8	17.2	17.9
Hoquiam, Wn.		B	7.9	11.1	12.3	14.9	16.4	17.8
		A	6.6	12.9	13.7	16.6	17.4	18.8
Seattle, Wn.		B	....	....	12.1	14.5	16.8	17.6
		A	....	10.4	13.5	15.6	17.0	17.8
Kalama, Wn.			3.8	9.4	10.6	13.5	14.3	....
Elyria, Ohio			6.0	9.0	10.0	14.7	16.0	17.5
Pittsburgh, Pa.			....	11.0	13.7	15.7	17.2	18.5
Trenton, N. J.			....	....	....	13.7	....	....
Ellensburg, Wn.			2.6	6.5	9.9	12.1	14.1	16.2
Duluth, Minn.		B	3.5	9.5	11.1	14.7	....	....
		A	6.3	11.0	13.8	15.8	....	....
Horace Mann	Nov. '16		5.9	11.2	12.1	15.1	12.6	15.3
(Columbia U.)	Apr. '17		9.7	12.7	11.2	16.4	16.6	17.0
N. Y. City School	Nov. '16		5.6	10.5	11.4	15.1	16.5	18.0
	Apr. '17		8.2	11.3	12.6	17.1	18.4	19.0
Denver, Colo.		B	5.1	9.8	9.9	12.8	13.5	16.2
		A	6.5	10.4	10.8	13.8	16.5	15.8
Idaho Springs, Colo.		B	....	....	7.3	9.4	12.9	15.2
	Average		5.9	10.4	11.6	14.6	15.9	17.2

DIVISION

CITY	DATE	SECTION	III	IV	V	VI	VII	VIII
Cherokee, Ia.			....	8.7	10.6	11.9	12.1	12.1
Webster City, Ia.			....	5.4	8.8	12.2	12.7	14.0
Hoquiam		B	....	9.1	9.9	11.1	12.6	13.1
		A	7.8	9.6	10.4	12.0	13.3	13.8
Seattle		B	....	....	7.6	10.5	12.0	12.6
		A	....	....	9.8	11.1	12.4	13.0
Kalama, Wn.			3.8	5.7	8.0	9.4	11.0	....
Elyria, Ohio			5.0	7.7	8.0	10.0	11.2	12.0
Pittsburgh, Pa.			....	7.0	9.8	11.5	12.9	13.9
Trenton, N. J.			....	....	....	9.8	....	....
Ellensburg, Wn.			1.8	4.0	5.9	8.9	9.8	11.7
Duluth		B	3.9	6.3	8.1	10.3	....	....
		A	4.6	7.2	9.2	11.1	....	....
Horace Mann	Nov. '16		4.1	7.2	8.9	11.0	12.2	11.8
(Columbia U.)	Apr. '17		6.3	10.0	8.1	12.5	11.8	12.9
N. Y. City School	Nov. '16		4.2	6.3	8.6	11.5	12.9	13.5
	Apr. '17		....	8.0	9.6	12.4	13.3	14.6
Denver, Colo.		B	4.2	6.8	7.8	9.8	10.7	11.2
		A	4.5	6.4	6.5	9.2	11.5	12.0
Idaho Springs		B	....	....	5.9	7.7	9.9	16.3
	Average		4.6	6.8	8.4	10.6	11.9	13.0

TABLE IX

Summary of Scores Attained in Woody Arithmetic Scales, Series A (Based upon Difficulty of Problems Solved)

Addition

City	Date	Section	III	IV	V	VI	VII	VIII
Cynthiana, Ky.	1916		4.50	6.22	7.16	7.27	7.44	7.98
Nassau County, N. Y.	1916		5.53	....	7.44	....	....	....
Paterson, N. J.	1918	B	5.03	....	6.93	....	....	....
.		A	5.16	....	7.18	....	....	....
Cities of Original Study	1915		4.99	6.11	6.99	7.95	8.65	9.01

Subtraction

City	Section	III	IV	V	VI	VII	VIII
Cynthiana, Ky.		3.54	4.32	5.59	6.26	6.20	7.57
Nassau County		4.40	....	5.78	....	....	....
Paterson	B	3.24	....	5.25	....	....	....
	A	3.71	....	5.56	....	....	....
Cities of Original Study		2.96	4.22	5.47	6.46	7.31	7.64

Multiplication

City	Section	III	IV	V	VI	VII	VIII
Cynthiana, Ky.		3.19	4.47	5.38	7.04	6.68	7.21
Nassau County		4.30	....	6.30	....	7.07	....
Paterson	B	....	....	....	....	6.94	....
	A	....	5.01	....	....	7.03	....
Cities of Original Study		1.89	4.05	5.53	6.72	7.26	7.93

Division

City	Section	III	IV	V	VI	VII	VIII
Cynthiana		....	3.76	4.91	5.96	5.72	6.62
Nassau County		3.18	....	5.36	....	6.64	....
Paterson	B	....	....	....	....	6.24	....
	A	....	4.01	....	....	6.59	....
Cities of Original Study		2.54	3.21	4.94	5.87	6.59	7.16

TABLE X

Summary of Scores Attained in Woody Arithmetic Scales, Series A (Based upon Median Number of Problems Solved)

Addition

City	Date	Section	III	IV	V	VI	VII	VIII
Wisconsin Cities	1916–17		15.5	20.2	22.7	28.4	31.9	33.1
Cities of Original Study	1915		14.5	18.3	23.1	29.8	32.4	34.0
Nassau County, N. Y.	1916		16.2	....	24.3	....	....	....
Seattle, Wn.	1916		....	....	....	29.2	31.4	32.4
Paterson, N. J.	1918	B	14.2	....	21.3	....	....	....
		A	16.0	....	23.0	....	....	....
Pittsburgh, Pa.	1919		14.5	18.3	23.1	29.7	32.4	33.9
Average			15.1	18.9	22.9	29.3	32.0	33.4

Subtraction

City	Section	III	IV	V	VI	VII	VIII
Wisconsin Cities		13.3	18.1	20.8	25.6	28.4	30.3
Cities of Original Study		11.2	15.7	20.4	25.0	28.5	31.7
Nassau County		16.9	....	23.1	....	....	....
Seattle		....	....	....	25.8	28.6	28.9
Paterson	B	12.5	....	19.5	....	....	....
	A	13.9	....	21.3	....	....	....
Pittsburgh		11.2	15.6	20.4	24.9	28.5	31.7
Average		13.0	16.5	20.9	25.3	28.5	30.8

Multiplication

City	Section	III	IV	V	VI	VII	VIII
Wisconsin Cities		6.8	15.2	19.2	27.3	30.9	33.2
Cities of Original Study		4.7	11.1	18.3	26.1	30.6	32.9
Nassau County		12.1	....	25.7	....	31.4	....
Seattle		....	....	....	27.6	30.0	37.0
Paterson	B	....	11.0	....	....	27.0	....
	A	....	15.2	....	....	29.1	....
Pittsburgh		5.8	9.8	16.4	23.7	27.4	30.1
Average		7.2	12.4	19.9	26.9	29.5	33.3

Division

City	Section	III	IV	V	VI	VII	VIII
Wisconsin Cities		7.6	13.5	19.6	25.1	28.4	30.0
Cities of Original Study		5.8	9.9	16.5	23.8	27.4	30.1
Nassau County		9.4	....	20.9	....	28.7	....
Seattle		....	....	....	24.2	26.5	26.1
Paterson	B	....	10.3	....	....	25.1	....
	A	....	13.1	....	....	26.9	....
Pittsburgh		5.8	9.8	16.4	27.4	27.4	38.0
Average		7.2	11.3	18.3	25.1	27.2	29.2

6. Presentation and Diagnosis of the Results

When the results for the different classes have been tabulated and the medians computed the findings should be set forth in appropriate tabular and graphic forms. For a single room the score card itself will suffice as a tabular representation, but in the larger school systems the score sheets from individual rooms are combined into building score sheets, showing the distribution and median achievement for each and every grade in the building, and these, in turn, into a city score sheet. Table XI represents the typical score sheet for a building or a city.

TABLE XI

Summary of Results in Addition, Series B, Ellensburg, Washington

	Grades			
Score	V	VI	VII	VIII
19			1	3
18			7	5
17		1	5	16
16		9	17	16
15		15	18	7
14	4	16	13	12
13	10	19	9	10
12	9	14	4	2
11	14	9	4	3
10	12	2		2
9	9	2		1
8	9	1	2	
7	2	1	1	
6	1			
5				
4	1		1	
3				
2				
1				
0				
No. of Pupils	71	89	82	77
Median	11.1	13.8	15.4	16.1

Such a table exhibits the class medians, the variability within each class distribution, and the great overlapping among the different class distributions.

The facts in this table will be more strikingly portrayed to both teachers and the public if presented graphically as in Fig. 1.

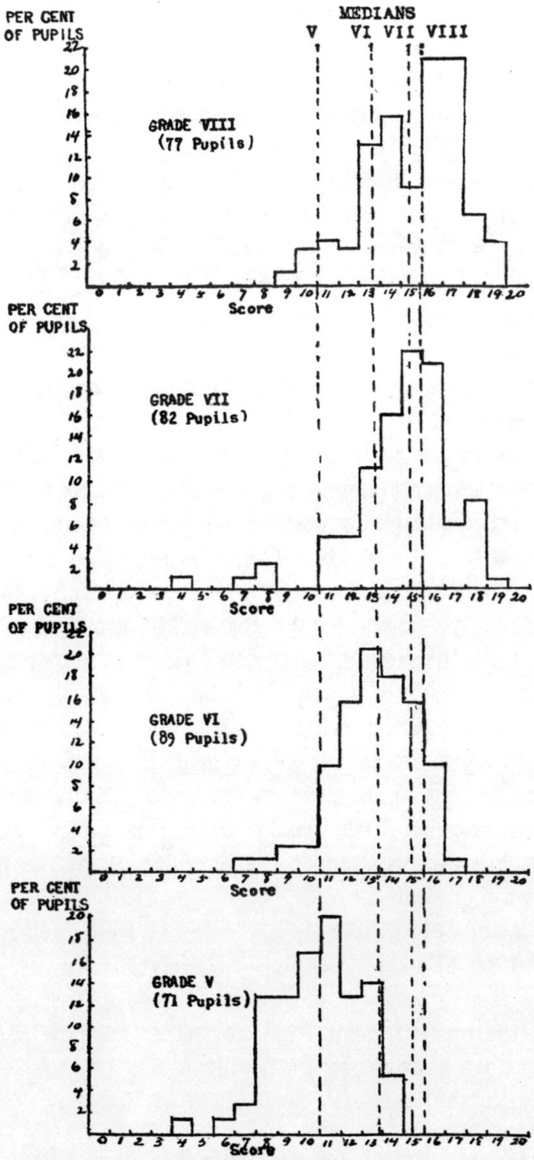

FIG. 1. DISTRIBUTION AND MEDIAN SCORES IN ADDITION, SERIES B
ELLENSBURG, WASH.

Such a graph is a composite, consisting of graphs for the different grades, drawn upon the same scale, placed one above the other in the proper order. In constructing these individual graphs the X axis represents the scale of possible scores, the Y axis, the scale for the percentage[1] of pupils making the various scores. The particular size of the units for these scales is arbitrary, depending upon the particular fact to be portrayed, and the size of the paper upon which the graph is to be made. The graph, as a rule, should show but one fact and this fact should be unmistakably clear.

From such a graph as Fig. 1, one can see at a glance the most common and median score in each distribution; the amount of progress from grade to grade as indicated by the distance between the grade medians, and the range of abilities within each grade. Closer examination will show the amount of overlapping. To illustrate, by adding the per cents indicated at the left of the fifth grade median line, it is seen that about 7 (6.8) per cent of the sixth grade, 5 (4.8) per cent of the seventh grade, and about 4 (3.9) per cent of the eighth grade do not reach the median of the fifth grade. On the other hand, by adding the various per cents to the right of the eighth grade median line, it is seen that none of the fifth grade children attain the median of the eighth grade, yet eleven per cent of the sixth grade and thirty-five per cent of the seventh grade do. Many other similar calculations can be made easily from this figure, and in no other way can the problem of the overlapping of the grades be so effectively presented.

On the whole, the graphs as presented in Fig. 1 tell more than any other type of graphic device. However, many other forms will prove useful in the interpretation and presentation of the results of the measurement. If one wishes to show the amount of progress from grade to grade through a comparison of the class medians, the form of graph given in Fig. 2 is very useful.

---

[1] Sometimes these graphs are made with the Y axis as a scale representing the number instead of the percentage of pupils making the various scores. This practice eliminates much computation. However, comparisons by per cents are apt to be more definite and more tangible.

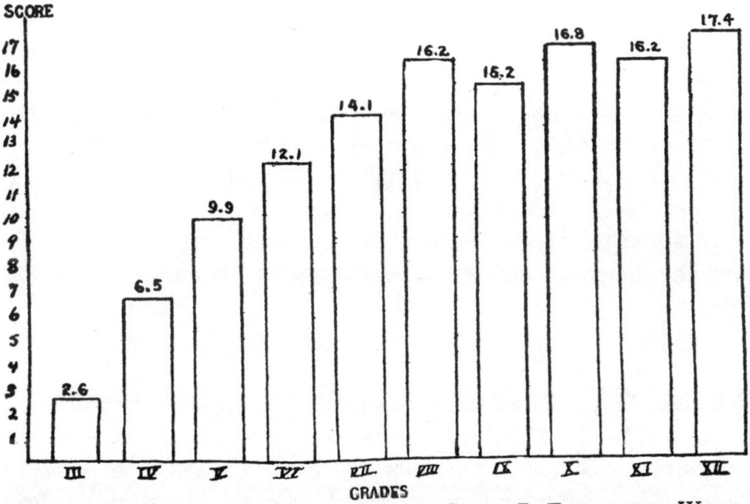

FIG. 2. MEDIAN SCORES IN MULTIPLICATION, SERIES B, ELLENSBURG, WASH.

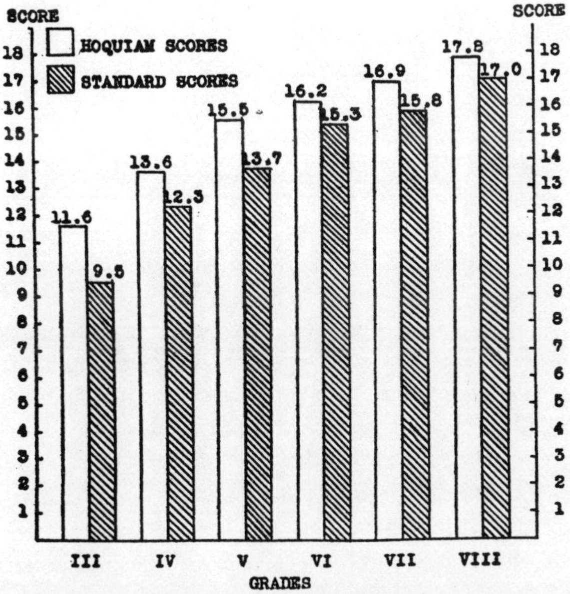

FIG. 3. COMPARISON OF MEDIAN SCORES (UPPER SECTIONS) IN ADDITION, SERIES B, WITH STANDARDS FOR EACH GRADE, HOQUIAM, WASH.

The comparative lengths of the rectangles clearly indicate the relative amounts of progress made and will single out the particular grades that need attention.

If one wishes to compare the median scores of certain grades with the *standards of achievement* for these grades, when setting forth the above facts, a graph similar to that in Fig. 3 is recommended. (Page 29.)

In this graph the unshaded rectangle represents the class medians for the A sections of the different grades (Hoquiam, Washington, 1917) in Addition, Series B, and the shaded area, the average of the scores attained in the different cities as set forth in Table VIII. Such a graph is very effective in determining whether the children in one school system are accomplishing as much as the children in other school systems, provided the two groups compared are comparable. Care must be taken to see that measurements made in one school system in the first months of the school year are not compared directly with the measurement made in some other system during the last months of the year. If possible, select the results from tests given at the same time of year, from the list of scores given in Tables VIII, IX and X. Otherwise, comparisons should be made from an estimate of what the standards should be at the time of giving the test, such estimate being based upon the list of given standards. The failure on the part of many school people to make legitimate comparisons with outside standards has convinced the writer that the greatest value of measurement is to come from comparison from measurements *inside* the system, and not outside.

The form of graph presented in Fig. 4 is especially recommended for use by the superintendent or principal in reporting the results to individual teachers where the teachers are especially sensitive concerning the results achieved.

The graph in solid black lines represents the distribution and median achievement of all the third grades in the city; the one in broken lines, the distribution and median achievement for an individual third grade room. Each of the teachers receives a copy of the solid line graph, and a broken line graph for her own room. The broken line graph is made upon tracing cloth

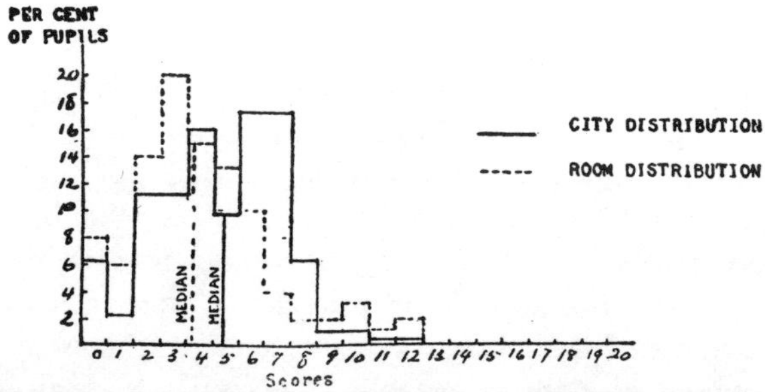

FIG. 4. DISTRIBUTION OF SCORES IN ADDITION, SERIES B, FOR A SINGLE THIRD GRADE ROOM SUPERIMPOSED ON THE THIRD GRADE DISTRIBUTION FOR CITY.

and can be superimposed upon the solid line graph. In this way the individual teacher can compare the results of her own pupils with the results achieved throughout the city, but she need not reveal the records achieved by her pupils unless she so desires.

While a single measurement with the Woody Arithmetic Scales, or any other tests for that matter, is very unreliable and unsafe, yet at times the teacher may wish to obtain a bird's eye view of the achievement of the individual pupils. The graph presented in Fig. 5 provides such a bird's eye view.

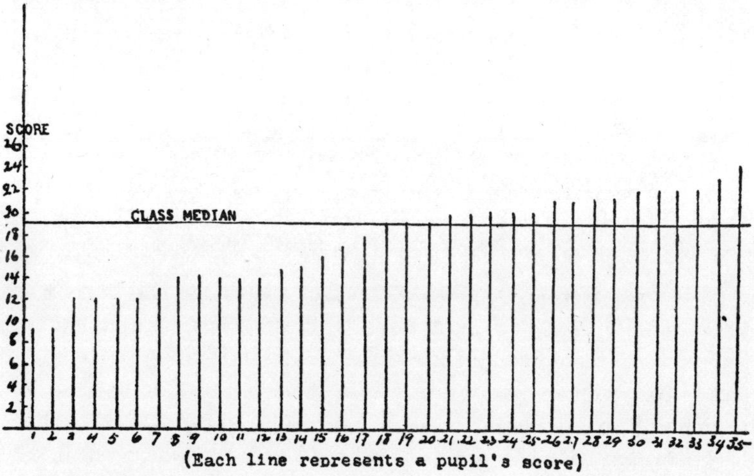

FIG. 5. INDIVIDUAL SCORES IN DIVISION, SERIES A, NORWALK, CONN.

In constructing such a graph, arrange the individual scores according to size and let a perpendicular line arising from the

X axis represent the score attained by an individual pupil, the size of the score determining the length of the line. From the appropriate position on the scale of scores on the Y axis, a line is drawn representing the median achievement for the class. A mere glance at this graph shows the relative standing of each and every pupil, and should suggest to the teacher that pupils with less than median achievement need more attention than those of more than median achievement. Of course, further testing should be done to establish the real achievement of the individual pupils, but the above graph will have much significance for the wide-awake teacher.

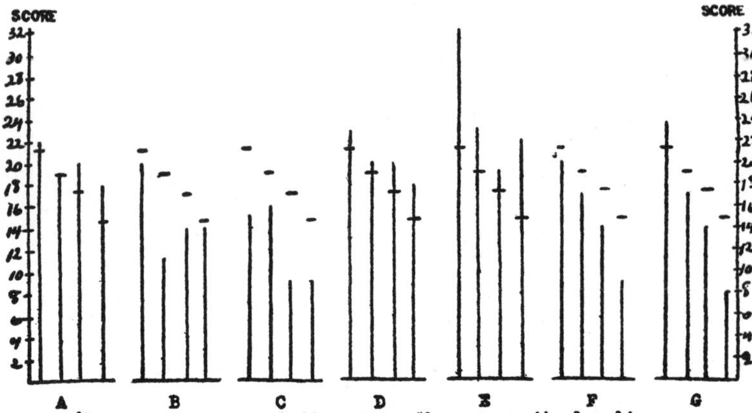

(Letters represent different pupils; perpendicular lines, pupil's score in addition, subtraction, multiplication and division respectively; cross bars, the class median)

FIG. 6. INDIVIDUAL SCORES COMPARED WITH ONE ANOTHER AND WITH THE CLASS MEDIAN. SERIES A, FIFTH GRADE, BRIDGETON, N. J.

If further testing establishes the fact that some children make exceptionally high or exceptionally low scores, one might be tempted to recommend that these particular children be promoted to a higher class or sent to a lower one. But before any promoting or demoting is done, special care should be taken to guarantee that the pupils concerned are good or poor in all phases of arithmetic. If the pupil's scores in the four fundamentals of arithmetic are graphed as in Fig. 6, the exact situation can be easily determined. For instance, individual A is above the class medians in all four processes, but not nearly so much as D and E; individuals B, C, and F are below the class medians; individual G is above the class median in addition, and below in the other three processes, being especially low in division. From suggestions gained from these graphs the teacher can direct more intelligently the drill work of these pupils, and when these facts are considered in relation to the class medians of the classes above and below, and the general character of the work in other phases of arithmetic, she can proceed to promote or demote more wisely.

Thus far the graphs drawn for the purpose of diagnosing the strengths or weaknesses of the system or of the individual have been based upon the total scores achieved; but real diagnosis, if it is to be helpful to a teacher must do more than consider mere total scores. Real diagnosis must take into consideration the types of problems missed and, if possible, ascertain the psychological process underlying the errors. Consequently, it is important to record the number of pupils solving each particular problem correctly. Such tabulation will reveal the types of problems causing a class special difficulty and will give possible suggestions concerning the individual. It is unsafe, however, to pass final judgment on the achievement of a class or an individual on success or failure in solving a single problem. In passing judgment upon the ability of a class to add the fundamental operations, all problems involving the single operations should be considered; to add fractions, all problems involving fractions. Many of the problems are so selected that they reveal ability to add the fundamental operations and ability to handle fractions, decimals, and denominate numbers at the same time, the particular ability or inability being determined from the nature of the answers.

The ability of a class, then, is determined by considering as many of the given problems as possible from as many angles as possible, and if the conclusions are to be reliable the class should be given other lists of similar problems. If this be true with regard to a class, it is doubly true with regard to an individual.

It has been intimated previously that real diagnosis consists in ascertaining the psychological process underlying

the errors in the solutions of the problems. It is interesting to know that many children are unable to add 17 even
2
—

though they can add much more difficult exercises involving more and harder combinations. It is not so significant to know that most children who miss the above exercise get the answer 10 as it is to know that the children get that answer by adding 7+2+1. These children have been absolutely accurate so far as adding the mere numbers are concerned, but they have misapplied a principle. The same thing is true with the children who add 23 and get the answer 514
25
16

or 91, the two most common errors with this exercise. Both answers are accurate as far as adding the mere numbers is concerned, but are inaccurate in the principle of carrying, the first answer being gotten by adding each column as a separate exercise and writing the answers so that they give the appearance of a single answer, while the second answer is gotten by putting down the 1 under the first column and carrying the 4. Misapplication of a principle causes children to subtract 11 and
7
—

get 16; 393 and get 225; 27 and get 15⅝.
178 — 12⅝ —

Many other illustrations could be cited showing real diagnosis, the discovery of the underlying psychological process causing success or failure; but the foregoing should be sufficient to demonstrate the value of studying the errors made. If the measurements are to be of greatest value to the classroom teacher a chart of the errors in each of the fundamental operations similar to the one in Fig. 7 should be made.[1]

Such a chart shows the teacher the particular things which should be emphasized in teaching division and indicates the method of realizing the greatest returns in teaching. It further suggests the possibility and value that come from analyzing

[1] Theisen, W. W. "The Diagnostic Value of the Woody Arithmetic Scales: A Reply." *Journal of Educational Psychology*, December, 1919.

	No.	%
1. Incorrect inverting or failure to invert	277	18.46
2. Incorrect Division	246	16.4
3. Incorrect placing of, or omitting of decimal point	189	12.6
4. Failure to reduce or incorrect reduction of remainder of answer	133	8.87
5. Incorrect division of denominate numbers	132	8.8
6. Omitting or adding cipher in quotient	101	6.73
7. Not completing division	92	6.13
8. Incorrect subtraction	81	5.4
9. Incorrect multiplication	61	4.07
10. Using incorrect denominator of remainder	36	2.4
11. Failure to classify result	35	2.33
12. Failure to employ both parts of fraction	29	1.93
13. Incorrect placing or bringing down	28	1.87
14. Incorrect reduction of mixed numbers	28	1.87
15. Failure to recognize the sign of operation	15	1.
16. Incorrect addition	7	.47
17. Error in copying	4	.27
18. Illegible answer	4	.27
19. Incorrect cancellation	2	.13
Total	1500	100%

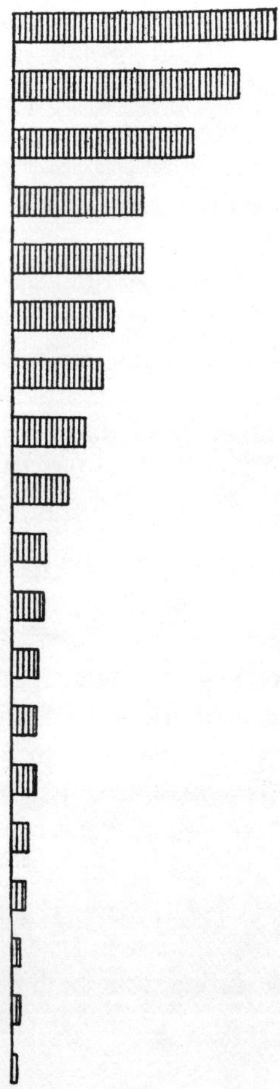

FIG. 7. ANALYSIS OF 1500 ERRORS IN DIVISION MADE BY GRADES VII AND VIII.

the causes of the particular responses which children make. In just this lies the hope for efficient teaching.

Other types of graphs could be suggested for interpreting and presenting the results of measurements, but it seems that the list suggested is adequate, the graphs therein presenting the situation in arithmetic with unmistakable clearness. Those making the measurements may feel that too many graphs have been recommended, but the full value of measurements cannot be realized until a thorough and extended analysis has been made. Until these analyses have been made, there is little basis for recommendations involving changes in subject matter, methods of teaching, time allotment or promotion, and such resulting recommendation will prove the worth of the measurement.

7. Supplementary Editions of the Scales

Parallel sets of scales are in process of construction which may be used to advantage in supplementing the original scales. They will be published by the Bureau of Publications, Teachers College.

A special arrangement of the original scales has been made by Dr. W. A. McCall, assistant professor of educational psychology, Teachers College, Columbia University. This is known as the Woody-McCall Mixed Fundamentals Tests and is different from the original scales in that each of the two equivalent test forms consists of problems selected from the addition, subtraction, multiplication, and division scales, the particular process involved being indicated for each problem. This edition has not been used very extensively, but it promises to be especially helpful in selecting especially capable children. Since this edition is rather unique Form I is reproduced here.

Mixed Fundamentals. Form I

Get the right answer to as many examples as you can in 20 minutes. Do all work on the front or back of this sheet.

(1) Add
2
3

(2) $2\times3=$

(3) $3\overline{)6}$

(4) Subtract
2
1

(5) Multiply
23
3

(6) Subtract
13
8

(7) Add
17
2

(8) $3+1=$

(9) Subtract
16
9

(10) Multiply
254
6

(11) $4\div2=$

(12) Add
23
25
16

(13) Subtract
393
178

(14) $2\overline{)13}$

(15) Add
9
24
12
15
19

(16) Multiply
5096
6

(17) $2\frac{3}{4} - 1=$

(18) Add
\$12.50
16.75
15.75

(19) Multiply
7898
9

(20) $\frac{1}{4}$ of $128=$

(21) Add
547
197
685
678
456
393
525
240
152

(22) Multiply
287
.05

(23) $248\div7=$

(24) Subtract
27
$12\frac{5}{8}$

(25) Add
4.0125
1.5907
4.10
8.673

(26) Multiply
9742
59

(27) $\frac{7}{8}$ of $624=$

(28) Add
.49
.28
.63
.95
1.69
.22
.33
.36
1.01
.56
.88
.75
.56
1.10
.18
.56

(29) $\frac{1}{8}\times2=$

(30) Multiply
$987\frac{3}{4}$
25

(31) $\frac{3}{4}\div5=$

(32) $7.3-3.00081=$

(33) 9)69 lb. 9 oz.

(34) Multiply
$.0963\frac{1}{8}$
.084

(35) $25.091+100.4+25+98.28+19.3614=$

[1] Copies of the Woody-McCall Mixed Fundamentals, Forms I and II, with direction sheet, may be obtained from the Bureau of Publications, Teachers College.

8. Supplementary References

Anderson, C. J.: "The Use of the Woody Scales for Diagnostic Purposes." *Elementary School Journal,* 17:770-781 (June 1918).

Monroe, Walter S.: "Experimental and Analytical Study of Woody's Arithmetical Scales, Series A." *School and Society,* 6:412-420 (October 6, 1917).

Theisen, W. W. and Fleming, Cecile White: "The Diagnostic Value of the Woody Arithmetic Scales. A Reply." *Journal of Educational Psychology,* Vol. IX, 475-488; 567-580 (November and December, 1918).

Theisen, W. W.: "A Report on the Use of Some Standard Tests for 1916-1917." Madison, Wisconsin, State Department of Public Instruction, 51-70, 1918.

See also *Research and Measurement Bulletin,* No. 2, January 1918. Pittsburgh, Pa.

PART II

THE DERIVATION OF THE SCALES

1. History of the Scales

The completed scales as shown in Part I of this monograph have been developed from about 20,000 test sheets. The first preliminary series of tests were given to a number of pupils in the public schools of Indiana and New Jersey. The preliminary series of tests consisted of a sheet of problems in addition and likewise one in subtraction, multiplication, and division. In constructing these preliminary lists there was a definite attempt to select problems of as great a variety as the fundamental processes would permit. There was also an attempt to begin the series in each process with the easiest problem that could be found and then gradually to increase the difficulty of each succeeding problem until the last ones in the series would be correctly solved by only a small percentage of the pupils in the eighth grade. By the selection of problems of such varied types and by giving the same lists of problems to pupils in all grades, it was thought that the diagnosing power of the lists would be greater and that the amount and the nature of the progress of one grade over another could best be determined.

The preliminary lists of problems in addition were given to 908 pupils, in subtraction to 916 pupils, in multiplication to 868 pupils, and in division to 696 pupils. The results of these preliminary lists showed that some of the problems were poorly selected and that they should be discarded. When the problems were ranked according to the total percentage of pupils solving them correctly, the results showed large gaps existing between the problems in particular portions of the series.

Guided by the results of these preliminary lists new lists were constructed. Only those problems of the original lists were chosen which were solved by a gradually increasing percentage of the pupils as one proceeded from the lower to the higher grades. If a problem were solved by a higher percentage of the pupils in the lower grades than in the higher grades it was rejected. Wherever there tended to be too large a step between two consecutive problems in the original series an attempt was made to interpose two or three problems of intermediate difficulty.

From the last week in October until the end of the second week in December, 1915, pupils were tested with these new lists of problems. These pupils were from seven different school systems located in Indiana, New Jersey, Connecticut, and New York. The addition problems were given to 4,489 pupils, the subtraction to 4,423 pupils, the multiplication to 3,922 pupils, and the division to 3,660 pupils. These pupils were distributed fairly equally from the second to the eighth grades inclusive.

All of the tests were given by the writer himself with the exception of those given to the pupils in two small school systems in Indiana.[1] The tests were given and the results scored according to the instructions given for administering the tests in Part I of this monograph with the one exception that no time limit for the solution of the problems was used. It was felt to be highly important, if the difficulty of each problem was to be firmly established, that each child should have a chance to solve each problem.

All of the tests were scored by the writer himself and thus the personal element in scoring was reduced to a minimum. The standard for marking a problem right or wrong as presented in Part I of this monograph was arbitrarily adopted. It was decided that a problem to be marked correct must be absolutely accurate and, wherever possible, reduced to its lowest terms. Otherwise, the problem was marked wrong. However, before adopting this arbitrary standard an effort was made to gain from teachers and supervisors of arithmetic the standards by which they marked a problem right or wrong. It was almost unanimously agreed that a problem must be absolutely accurate and reduced to its lowest terms. Thus the arbitrary standard adopted by the writer is in accordance with the best practice exercised in the teaching of arithmetic.

The results of these tests were recorded in two ways:

1. The pupils were distributed according to the number of

[1] Those giving the tests in these two systems were men who have had experience in giving tests and who could be trusted to carry out the writer's directions.

TABLE XII

DISTRIBUTION ACCORDING TO THE NUMBER OF ADDITION PROBLEMS SOLVED

	GRADE II	GRADE III	GRADE IV	GRADE V	GRADE VI	GRADE VII	GRADE VIII
38					3	21	41
37					15	37	33
36					30	82	55
35				2	37	96	72
34				4	45	91	70
33				1	51	90	76
32				8	34	75	46
31				13	45	83	45
30			1	13	45	49	27
29			0	26	51	57	20
28			2	35	33	53	18
27			1	32	36	48	19
26		1	2	46	37	34	10
25		0	11	40	34	34	4
24		0	5	54	37	16	4
23		3	33	75	29	27	1
22		6	47	64	25	8	2
21	1	11	42	77	15	7	1
20	0	10	54	54	15	4	
19	0	26	65	49	6	3	
18	0	43	56	43	5	2	
17	4	47	75	28	3		
16	3	64	72	10	2		
15	7	70	42	7			
14	7	54	24	3			
13	13	44	14	0			
12	10	40	18	1			
11	43	39	16	1			
10	31	33	7	0			
9	46	35	5	0			
8	38	23	2	0			
7	35	16	1	0			
6	36	10	2	0			
5	69	14	1	1			
4	48	8	2				
3	43	4	1				
2	17	4	0				
1	13	6	1				
0	25	4					
No. Tested. .	489	615	602	687	633	917	544
Median	6.819	14.509	18.321	23.073	29.774	32.446	33.987
25 per cent. .	4.505	10.902	16.201	20.532	25.625	28.872	31.667
75 " " . .	9.929	16.894	20.694	26.206	33.446	35.070	35.903
Quartile.	2.712	2.996	2.247	2.837	3.910	3.099	2.118

TABLE XIII

NUMBER IN EACH GRADE THAT SOLVED EACH PROBLEM IN ADDITION CORRECTLY

PROBLEM NO.	GRADE II	GRADE III	GRADE IV	GRADE V	GRADE VI	GRADE VII	GRADE VIII
1	388	456	499	654	622	896	541
2	433	582	595	681	630	913	542
3	392	593	595	680	626	911	539
4	326	468	521	659	614	901	540
5	323	501	554	673	629	914	544
6	279	530	565	668	628	911	542
7	259	538	565	679	631	915	542
8	220	474	542	665	624	911	544
9	165	530	568	667	613	880	522
10	152	531	570	663	623	895	539
11	190	543	577	663	608	886	535
12	52	399	541	657	620	896	537
13	32	229	373	627	622	901	537
14	37	405	541	664	627	900	534
15	23	328	499	627	602	876	530
16	6	238	387	567	533	806	500
17	22	288	431	551	500	787	475
18	8	208	386	539	511	801	505
19	1	92	246	399	436	662	457
20	1	87	307	555	586	883	528
21	1	71	276	498	564	839	498
22		49	204	441	528	814	489
23		4	34	308	490	771	500
24		0	4	99	296	521	385
25		3	14	213	397	651	457
26		2	11	192	423	682	483
27		0	10	178	369	678	470
28		0	14	166	414	693	448
29		0	8	131	300	591	409
30		2	33	157	403	684	462
31		3	34	164	317	490	344
32		0	4	157	373	674	421
33		0	3	57	235	461	290
34		15	128	271	338	684	432
35		4	57	169	318	558	392
36		0	2	40	179	537	354
37		0	1	20	176	529	359
38		0	1	9	155	240	274
No. Tested.	489	615	602	687	633	917	544

problems solved correctly. Table XII represents the distribution for the problems in addition. Beginning at the lower left-hand corner, the table shows that 25 out of 489 pupils in the second grade, and 4 out of 615 pupils in the third grade were unable to solve a single problem, etc. This table also shows the median achievement of each grade distribution. The median achievement of a class is such a number of problems correctly solved that there are just as many pupils who solve a greater number of problems as there are those who solve a smaller number. This table shows the range in the number of problems correctly solved that will include the middle 50 per cent of the pupils. It also shows the variability in terms of the quartile, or, as it is sometimes designated, the "semi-inter-quartile range."

2. The results were tabulated in another method so as to record the number of pupils who solved each individual problem correctly. Thus Table XIII shows that 388 out of 489 pupils in the second grade solved problem No. 1; 433 pupils solved problem No. 2, etc. From these two crude summaries given in Tables XII and XIII the addition scales have been developed.[1]

2. P.E. as a Unit of Measure

It may be said that we have always measured pupils in the fundamental operations of arithmetic. It may be said that schools and school systems have likewise been measured. No doubt this is true. Whenever a teacher says that one boy is better in addition than another boy, in a certain sense, she measures him. Whenever we compare one individual with another individual, one quality with another quality, or one class with another class, we are measuring. Such standards of measurements as these are no doubt inaccurate and changeable. Whenever a teacher measures a class by means of an examination she tends to have a more constant and more objective measurement. The relation of the different questions of the exam-

[1] Similar tables for the problems in subtraction, multiplication, and division will be found at the end of Part II. In the discussion of the derivation of the scales I shall show in detail the method by which the scale in addition was developed, and shall not discuss the other processes. However, I shall include the final values of each problem in each of the other processes and the most important tables of crude data from which the established values were determined.

ination to one another, however, is unknown. All the questions may be of equal difficulty, or one may be several times as difficult as another. The chief value of a scale as a means of measurement is that it is made up of a number of distinct units whose value is known and remains constant. Such a scale can be used by different people in making similar measurement and the results will be comparable. On the linear rule the unit of measurement is the inch or centimeter; on the thermometer, the degree. Everyone knows what is meant when we speak of an inch, a degree, or any fractional part thereof. These amounts are very definite and always have the same meaning. Moreover almost any one can make reliable measurements with a rule or with a thermometer.

In the building of these arithmetic scales there has been a definite attempt to approximate as closely as possible the accuracy and the constancy of the ruler or the thermometer. The difficulty of each problem has been established and its position above a selected zero point determined. The problems have all been placed in their relative positions on a projected linear scale. In the scales of Series B a definite attempt has been made to select problems with equal amounts of difficulty between them. The unit of measure of difficulty on these arithmetic scales, which corresponds to the inch on the ruler or to the degree on the thermometer, is what is called in statistical terms the Median Deviation or Probable Error. (P.E.)

Before taking up the significance of the median deviation let us discuss the normal surface of frequency. In the construction of these scales, it has been assumed that achievement in the solution of problems in the fundamental processes is distributed according to the normal surface of frequency. Furthermore it has been assumed that the variability of any grade from the second to the eighth is equal to that of any other.

These assumptions are based upon the well-established principle that intellectual abilities are distributed in the same way as are physical traits. If we should arrange one thousand men, selected at random, in a row according to their height, we should find a very large group of men in the center who are about medium height. On one end of the row would be a few very short men and on the other end would be a few very tall men. Likewise if we assume that achievement in the solution of problems in the fundamental processes in any grade is distributed normally, then we should expect to find a large number of the class solving about the same number of problems; furthermore we should expect to find a few dull pupils who can solve but just a few problems and a few bright pupils who can solve more than the average number of problems. The so-called normal curve illustrating such a distribution is reproduced in Fig. 8. The properties of the normal curve have been most accurately determined. Let us assume that Fig. 8 represents the achievement in the solution of problems among a large number of third grade pupils.

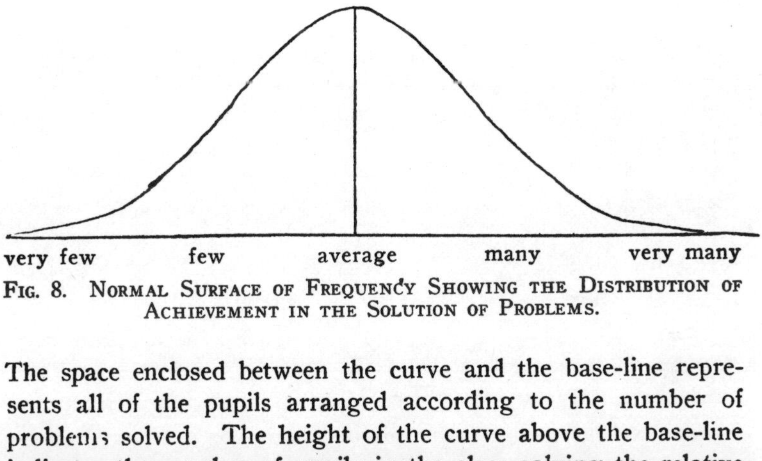

FIG. 8. NORMAL SURFACE OF FREQUENCY SHOWING THE DISTRIBUTION OF ACHIEVEMENT IN THE SOLUTION OF PROBLEMS.

The space enclosed between the curve and the base-line represents all of the pupils arranged according to the number of problems solved. The height of the curve above the base-line indicates the number of pupils in the class solving the relative number of problems shown on the base-line. Each pupil is represented by an equal amount of the enclosed area. Thus, at the extreme left the curve is very near the base, which indicates the small number of pupils who were able to solve only a very few problems. In the middle the curve is distant from the base-line representing the large number of pupils who solved an average number of problems; at the extreme right the curve is very near the base, which indicates the small number of pupils who are able to solve many more than the average number of problems.

If our assumption with regard to the achievement in the solution of problems is true, then the graphic representations of the tables of distribution according to the number of problems solved must be similar to Fig. 8.

Figs. 9 to 15 inclusive represent graphically the distribution of the achievement in the solution of the addition problems throughout the various grades. These figures on the whole correspond fairly well to the normal curve of distribution. It will be seen that in the second grade distribution the curve is somewhat skewed to the left. This is probably due to the fact that a great number of the teachers were just beginning to teach the fundamental operations to their classes. It will also be seen that the distributions for grades seven and eight are skewed somewhat to the right. This indicates the need for one or two more difficult problems at the end of the addition series. It will be noted from the distribution tables in the back of this monograph that the distributions for the other processes conform to the normal curve better than the foregoing figures.

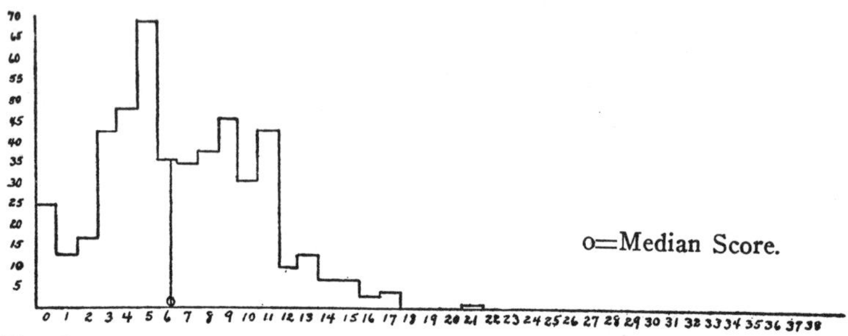

FIG. 9. DISTRIBUTION ACCORDING TO THE NUMBER OF ADDITION PROBLEMS SOLVED IN SECOND GRADE.

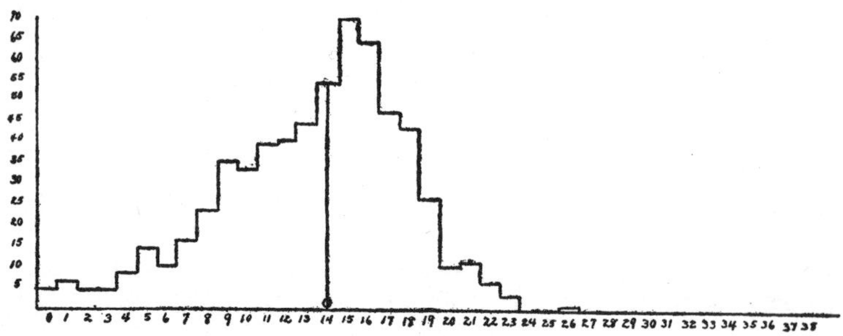

FIG. 10. DISTRIBUTION ACCORDING TO THE NUMBER OF ADDITION PROBLEMS SOLVED IN THIRD GRADE.

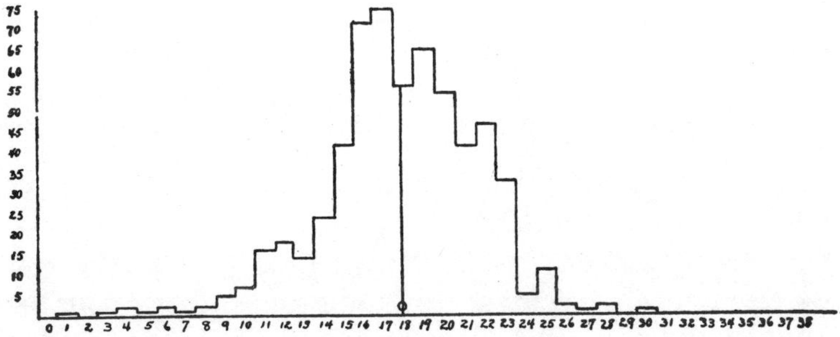

FIG. 11. DISTRIBUTION ACCORDING TO THE NUMBER OF ADDITION PROBLEMS SOLVED IN FOURTH GRADE.

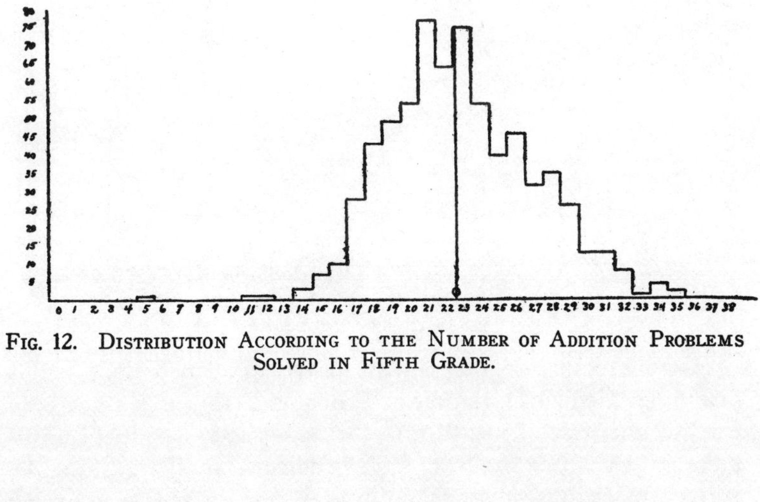

FIG. 12. DISTRIBUTION ACCORDING TO THE NUMBER OF ADDITION PROBLEMS SOLVED IN FIFTH GRADE.

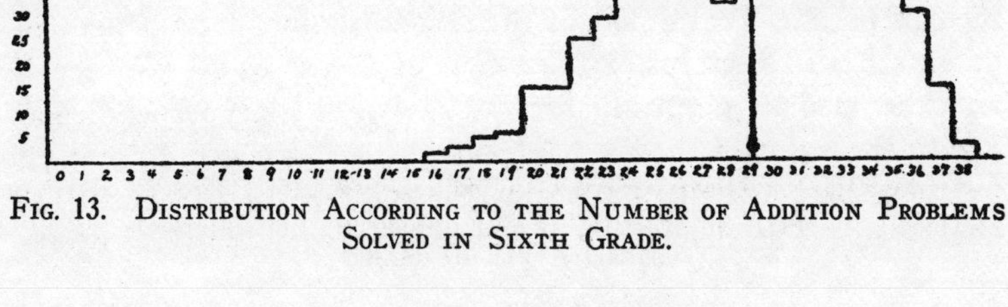

FIG. 13. DISTRIBUTION ACCORDING TO THE NUMBER OF ADDITION PROBLEMS SOLVED IN SIXTH GRADE.

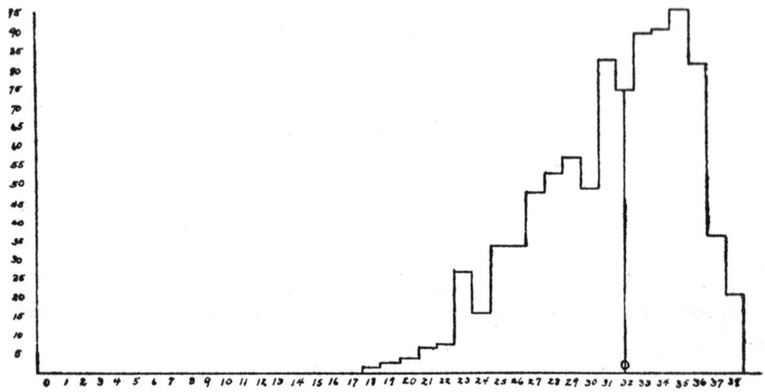

FIG. 14. DISTRIBUTION ACCORDING TO THE NUMBER OF ADDITION PROBLEMS SOLVED IN SEVENTH GRADE.

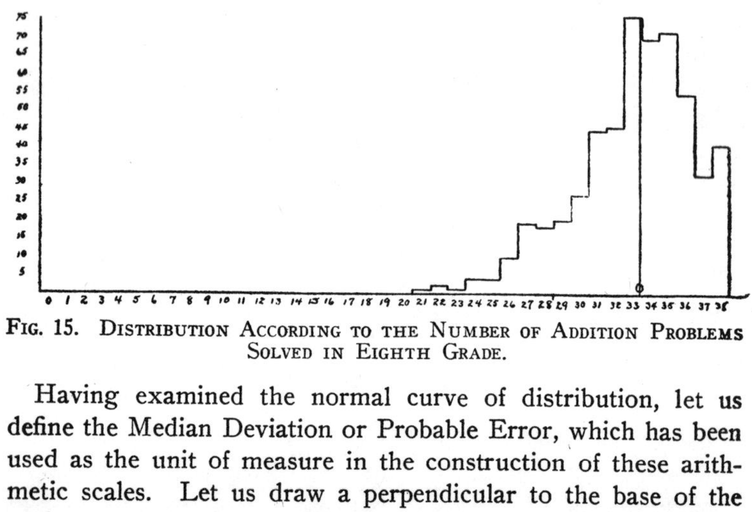

FIG. 15. DISTRIBUTION ACCORDING TO THE NUMBER OF ADDITION PROBLEMS SOLVED IN EIGHTH GRADE.

Having examined the normal curve of distribution, let us define the Median Deviation or Probable Error, which has been used as the unit of measure in the construction of these arithmetic scales. Let us draw a perpendicular to the base of the surface of frequency so that fifty per cent of all the cases lie on one side of the perpendicular and fifty per cent on the other side. The point where the perpendicular cuts the base is the median point. To the left of the median point, draw a perpendicular *f a* so that just 25 per cent of the cases lie between it and the median perpendicular. Draw a similar perpendicular *d c* to the right of the median point. The area *a c d f* cut off by these perpendiculars contains the middle 50 per cent of all the cases. The distance *a m* or *m c* on the base-line of the

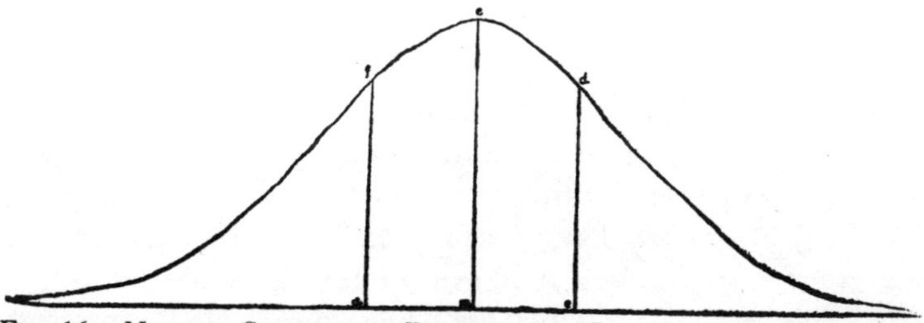

FIG. 16. NORMAL SURFACE OF DISTRIBUTION SHOWING THE MEDIAN AND P.E. DISTANCE AT EACH SIDE OF THE MEDIAN POINT.

surface of frequency is the Median Deviation or the Probable Error. The Probable Error or P.E., as it will be called throughout this monograph, is thus the distance along the base-line of a surface of distribution from the median point to the perpendicular on either side of the median which cuts off 25 per cent of the cases.

Furthermore, it has been established that 2 P.E. is the distance from the median point to the perpendicular on either side of the median which cuts off 41.13 per cent of the cases; 3 P.E., the distance which cuts off 47.85 per cent of the cases; and 4 P.E., the distance which cuts off 49.65 per cent of the cases. Theoretically the curve and the base-line never meet but continually approach one another as the distance from the median point increases. For the purposes of this study we may consider that they meet at a distance of 4.6 P.E. from the median point, for a perpendicular erected here on either side of the median cuts off but 0.1 per cent of all the cases. These facts enable us to locate each problem in its proper position on the base of any grade distribution.

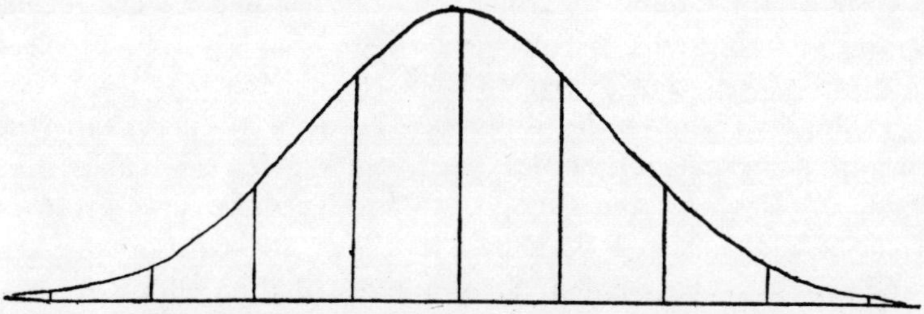

FIG. 17. NORMAL SURFACE OF FREQUENCY SHOWING P.E. DISTANCES FROM THE MEDIAN POINT.

3. Scaling the Problems in Addition for Each Grade

Since we have assumed that achievement in the solution of problems in the fundamental processes is distributed according to the normal surface of frequency and since we have adopted the P.E. of a grade distribution as the unit of measurement, it is an easy matter to locate each problem on the base-line of each grade distribution. It is evident that a problem which is solved by exactly 50 per cent of pupils in any class represents the median achievement of the class and that it would be located at the median point of the base-line. By definition, P.E. is the distance along the base-line from the median point to the perpendicular on either side of the median which cuts off 25 per cent of the cases. Evidently then a problem that is solved by 75 per cent of the pupils would be 1 P.E. too easy to represent the median achievement of the class and would be located at −1 P.E. distance from the median point. Likewise a problem that is solved by only 25 per cent of the pupils is too difficult to represent the median achievement and would be located at + 1 P.E. distance from the median point. Thus, if we know what per cent of a class solved any problem, it is easy to find the deviation of this per cent from 50 per cent or the median achievement of the class. If this per cent of deviation from the median achievement is known in terms of P.E., we can locate any problem with reference to the median of that distribution. Table XIV gives the P.E. value for each tenth of a per cent deviation from the median point of a normal distribution (i.e., deviation of 0.0 per cent to 49.9 per cent above or below the median).[1]

Table XIII previously given (page 42) shows the number in each grade that solved each problem in addition correctly. Table XV shows these numbers reduced into terms of per cents. Thus in the second grade 79.4 per cent of the pupils solved problem No. 1; 88.6 per cent solved problem No. 2, etc.

Table XVI shows the difference between 50 per cent (the median achievement) and the per cents given in Table XV. Table XVII shows the P.E. values for the differences given in Table XVI. These P.E. values represent the position of each problem on the base-line of the grade distribution with reference

[1] Table XIV is taken directly from B. R. Buckingham's *Spelling Ability* (Table XLVII). It is a modification of the table given in E. L. Thorndike's *Mental and Social Measurements* (page 200).

TABLE XIV

P.E. VALUES CORRESPONDING TO GIVEN PER CENTS OF THE NORMAL SURFACE OF FREQUENCY, PER CENTS BEING TAKEN FROM THE MEDIAN

%	.0	.1	.2	.3	.4	.5	.6	.7	.8	.9
0	.000	.004	.007	.011	.015	.019	.022	.026	.030	.033
1	.037	.041	.044	.048	.052	.056	.059	.063	.067	.071
2	.074	.078	.082	.085	.089	.093	.097	.100	.104	.108
3	.112	.115	.119	.123	.127	.130	.134	.138	.141	.145
4	.149	.153	.156	.160	.164	.168	.172	.175	.179	.183
5	.187	.190	.194	.198	.201	.205	.209	.213	.216	.220
6	.224	.228	.231	.235	.239	.243	.246	.250	.254	.258
7	.261	.265	.269	.273	.277	.280	.284	.288	.292	.296
8	.299	.303	.307	.311	.315	.318	.322	.326	.330	.334
9	.337	.341	.345	.349	.353	.357	.360	.364	.368	.372
10	.376	.380	.383	.387	.391	.395	.399	.403	.407	.410
11	.414	.418	.422	.426	.430	.434	.437	.441	.445	.449
12	.453	.457	.461	.464	.468	.472	.476	.480	.484	.489
13	.492	.496	.500	.504	.508	.512	.516	.519	.523	.527
14	.531	.535	.539	.543	.547	.551	.555	.559	.563	.567
15	.571	.575	.579	.583	.588	.592	.596	.600	.603	.608
16	.612	.616	.620	.624	.628	.632	.636	.640	.644	.648
17	.652	.656	.660	.665	.669	.673	.677	.681	.685	.689
18	.693	.698	.702	.706	.710	.714	.719	.723	.727	.731
19	.735	.740	.744	.748	.752	.756	.761	.765	.769	.773
20	.778	.782	.786	.790	.795	.799	.803	.807	.812	.816
21	.820	.825	.829	.834	.838	.842	.847	.851	.855	.860
22	.864	.869	.873	.878	.882	.886	.891	.895	.900	.904
23	.909	.913	.918	.922	.927	.931	.936	.940	.945	.949
24	.954	.958	.963	.968	.972	.977	.982	.986	.991	.996
25	1.000	1.005	1.009	1.014	1.019	1.024	1.028	1.033	1.038	1.042
26	1.047	1.052	1.057	1.062	1.067	1.071	1.076	1.081	1.086	1.091
27	1.096	1.101	1.105	1.110	1.115	1.120	1.125	1.130	1.135	1.140
28	1.145	1.150	1.155	1.160	1.165	1.170	1.176	1.181	1.186	1.191
29	1.196	1.201	1.206	1.211	1.217	1.222	1.227	1.232	1.238	1.243
30	1.248	1.253	1.259	1.264	1.269	1.275	1.279	1.286	1.291	1.296
31	1.302	1.307	1.313	1.318	1.324	1.329	1.335	1.340	1.346	1.351
32	1.357	1.363	1.368	1.374	1.380	1.386	1.391	1.397	1.403	1.409
33	1.415	1.421	1.427	1.432	1.438	1.444	1.450	1.456	1.462	1.469
34	1.475	1.481	1.487	1.493	1.499	1.506	1.512	1.518	1.524	1.531
35	1.537	1.543	1.549	1.556	1.563	1.569	1.576	1.582	1.589	1.595
36	1.602	1.609	1.616	1.622	1.629	1.636	1.643	1.649	1.656	1.663
37	1.670	1.677	1.685	1.692	1.699	1.706	1.713	1.720	1.728	1.735
38	1.742	1.749	1.757	1.765	1.772	1.780	1.788	1.795	1.803	1.811
39	1.819	1.827	1.835	1.843	1.851	1.859	1.867	1.875	1.884	1.892
41	1.900	1.909	1.918	1.926	1.935	1.944	1.953	1.962	1.971	1.979
42	1.988	1.997	2.007	2.016	2.026	2.035	2.044	2.054	2.064	2.074
03	2.083	2.093	2.103	2.114	2.124	2.134	2.145	2.155	2.166	2.177
44	2.188	2.199	2.211	2.222	2.234	2.245	2.257	2.269	2.281	2.293
44	2.305	2.318	2.331	2.344	2.357	2.370	2.384	2.397	2.411	2.425
45	2.439	2.453	2.468	2.483	2.498	2.514	2.530	2.546	2.562	2.579
46	2.597	2.614	2.631	2.648	2.667	2.686	2.706	2.726	2.746	2.767
47	2.789	2.811	2.834	2.857	2.881	2.905	2.932	2.958	2.986	3.015
48	3.044	3.077	3.111	3.146	3.182	3.219	3.258	3.300	3.346	3.395
49	3.450	3.506	3.571	3.643	3.725	3.820	3.938	4.083	4.275	4.600
50										

TABLE XV

PER CENT IN EACH GRADE THAT SOLVED EACH PROBLEM IN ADDITION CORRECTLY

PROBLEM NO.	GRADE II	GRADE III	GRADE IV	GRADE V	GRADE VI	GRADE VII	GRADE VIII
1	79.4	74.2	82.9	95.2	98.3	97.7	99.5
2	88.6	94.6	98.9	99.2	99.6	99.6	99.7
3	80.2	96.4	98.9	99.0	98.9	99.4	99.1
4	66.7	76.1	86.6	96.0	97.1	98.3	99.3
5	66.1	81.5	92.1	98.0	99.4	99.7	100.0
6	57.1	86.2	93.9	97.2	99.2	99.4	99.7
7	53.0	87.5	93.9	98.8	99.7	99.8	99.7
8	45.0	77.1	90.1	96.8	98.6	99.4	100.0
9	33.8	86.2	94.4	97.1	96.9	96.0	96.0
10	31.1	86.4	94.7	96.5	98.4	97.6	99.1
11	38.9	88.3	95.9	96.5	96.1	96.6	98.4
12	10.7	64.9	89.9	95.7	97.9	97.7	98.7
13	6.6	37.3	62.0	91.3	98.3	98.3	98.7
14	7.6	65.9	89.9	96.7	99.1	98.2	98.2
15	4.7	53.3	82.9	91.3	95.1	95.5	97.5
16	1.2	38.7	64.3	82.6	84.2	87.9	92.0
17	4.5	46.8	71.6	80.2	79.1	85.8	87.4
18	1.6	33.8	64.2	78.4	80.8	87.4	92.9
19	.2	15.0	40.9	58.1	68.9	72.2	84.0
20	.2	14.2	51.0	80.8	92.6	96.3	97.1
21	.2	11.6	45.9	72.5	89.1	91.5	91.6
22		8.0	33.9	64.2	83.4	88.8	89.9
23		.6	5.6	44.9	77.4	84.1	92.0
24			.6	14.4	46.8	56.8	70.7
25		.5	2.3	31.1	62.7	71.0	84.0
26		.3	1.8	28.0	66.9	74.4	88.8
27			1.6	26.0	58.3	74.0	86.4
28			2.3	24.2	65.4	75.6	82.4
29			1.3	19.1	47.4	64.5	75.2
30		.3	5.5	22.9	63.7	74.6	85.0
31		.5	5.6	23.9	50.1	53.4	63.3
32			.6	22.9	58.9	73.5	77.4
33			.5	8.3	37.1	50.3	53.4
34		2.4	21.3	39.5	53.4	74.6	79.5
35		.6	9.5	24.6	50.3	60.9	72.1
36			.3	5.8	28.3	58.6	65.1
37			.2	2.9	27.9	57.7	66.0
38			.2	1.3	24.5	26.2	50.4

to the median point. These values enable us to scale the problems.

By reference to Table XV it is seen that problem No. 1 was solved by 79.4 per cent of the pupils in the second grade. Table XVI shows a difference of 29.4 per cent between the median achievement (i.e., a problem solved by 50 per cent of the class)

TABLE XVI

DIFFERENCE BETWEEN FIFTY PER CENT AND THE PER CENT IN EACH GRADE THAT SOLVED EACH PROBLEM IN ADDITION CORRECTLY

PROBLEM NO.	GRADE II	GRADE III	GRADE IV	GRADE V	GRADE VI	GRADE VII	GRADE VIII
1	29.4	24.2	32.9	45.2	48.3	47.7	49.5
2	38.6	44.6	48.9	49.2	49.6	49.6	49.7
3	30.2	46.4	48.9	49.0	48.9	49.4	49.1
4	16.7	26.1	36.6	46.0	47.1	48.3	49.3
5	16.1	31.5	42.1	48.0	49.4	49.7	50.0
6	7.1	36.2	43.9	47.2	49.2	49.4	49.7
7	3.0	37.5	43.9	48.8	49.7	49.8	49.7
8	—5.0	27.1	40.1	46.8	48.6	49.4	50.0
9	—16.2	36.2	44.4	47.1	46.9	46.0	46.0
10	—18.9	36.4	44.7	46.5	48.4	47.6	49.1
11	—11.1	38.3	45.9	46.5	46.1	46.6	48.4
12	—39.3	14.9	39.9	45.7	47.9	47.7	48.7
13	—43.4	—12.7	12.0	41.3	48.3	48.3	48.7
14	—42.4	15.9	39.9	46.7	49.1	48.2	48.2
15	—45.3	3.3	32.9	41.3	45.1	45.5	47.5
16	—48.8	—11.3	14.3	32.6	38.2	37.9	42.0
17	—45.5	—3.2	21.6	30.2	29.1	35.8	37.4
18	—48.4	—16.2	14.2	28.4	30.8	37.4	42.9
19	—49.8	—35.0	—9.1	8.1	18.9	22.2	34.0
20	—49.8	—35.8	1.0	30.8	42.6	46.3	47.1
21	—49.8	—38.4	—4.1	22.5	39.1	41.5	41.6
22		—42.0	—16.1	14.2	33.4	38.8	39.9
23		—49.4	—44.4	—5.1	27.4	34.1	42.0
24		—50.0	—49.4	—35.6	—3.2	6.8	20.7
25		—49.5	—47.7	—18.9	12.7	21.0	34.0
26		—49.7	—48.2	—22.0	16.9	24.4	38.8
27		—50.0	—48.4	—24.0	8.3	24.0	36.4
28		—50.0	—47.7	—25.8	15.4	25.6	32.4
29		—50.0	—48.7	—30.9	—2.6	14.5	25.2
30		—49.7	—44.5	—27.1	13.7	24.6	35.0
31		—49.5	—44.4	—26.1	.1	3.4	13.3
32		—50.0	—49.4	—27.1	8.9	23.5	27.4
33		—50.0	—49.5	—41.7	—12.9	.3	3.4
34		—47.6	—28.7	—10.5	3.4	24.6	29.5
35		—49.4	—40.5	—25.4	.3	10.9	22.1
36		—50.0	—49.7	—44.2	—21.7	8.6	15.1
37		—50.0	—49.8	—47.1	—22.1	7.7	16.0
38		—50.0	—49.8	—48.7	—25.5	—23.8	.4

and the per cent that actually solved this problem. Table XVII shows that this problem would be located at -1.217 P.E. from the median point in the second grade distribution. Table XVII shows that this same problem would be located at -.963 P.E. from the median of the third grade distribution, etc. Table

TABLE XVII

P.E. EQUIVALENT OF DIFFERENCE BETWEEN FIFTY PER CENT AND THE PER CENT IN EACH GRADE THAT SOLVED EACH PROBLEM IN ADDITION CORRECTLY

PROBLEM NO.	GRADE II	GRADE III	GRADE IV	GRADE V	GRADE VI	GRADE VII	GRADE VIII
1	—1.217	— .963	—1.409	—2.468	—3.146	—2.958	—3.820
2	—1.788	—2.384	—3.395	—3.571	—3.938	—3.938	—4.083
3	—1.259	—2.667	—3.395	—3.450	—3.395	—3.725	—3.506
4	— .640	—1.052	—1.643	—2.597	—2.811	—3.146	—3.643
5	— .616	—1.329	—2.093	—3.044	—3.725	—4.083	—4.600
6	— .265	—1.616	—2.293	—2.834	—3.571	—3.725	—4.083
7	— .112	—1.706	—2.293	—3.346	—4.083	—4.275	—4.083
8	.187	—1.101	—1.909	—2.746	—3.258	—3.725	—4.600
9	.620	—1.616	—2.357	—2.811	—2.767	—2.597	—2.597
10	.731	—1.629	—2.397	—2.686	—3.182	—2.932	—3.506
11	.418	—1.765	—2.579	—2.686	—2.614	—2.706	—3.182
12	1.843	— .567	—1.892	—2.546	—3.015	—2.958	—3.300
13	2.234	.480	— .453	—2.016	—3.146	—3.146	—3.300
14	2.124	— .608	—1.892	—2.726	—3.506	—3.111	—3.111
15	2.483	— .123	—1.409	—2.016	—2.453	—2.514	—2.905
16	3.346	.426	— .543	—1.391	—1.757	—1.735	—2.083
17	2.514	.119	— .847	—1.259	—1.201	—1.589	—1.699
18	3.182	.620	— .539	—1.165	—1.291	—1.669	—2.177
19	4.275	1.537	.341	— .303	— .731	— .873	—1.475
20	4.275	1.589	— .037	—1.291	—2.145	—2.648	—2.811
21	4.275	1.772	.153	— .886	—1.827	—2.035	—2.044
22		2.083	.616	— .539	—1.438	—1.803	—1.892
23		3.725	2.357	.190	—1.115	—1.481	—2.083
24			3.725	1.576	.119	— .254	— .807
25		3.820	2.958	.731	— .480	— .820	—1.475
26		4.083	3.111	.864	— .648	— .972	—1.803
27			3.182	.954	— .311	— .954	—1.629
28			2.958	1.038	— .588	—1.028	—1.380
29			3.300	1.296	.097	— .551	—1.009
30		4.083	2.370	1.101	— .519	— .982	—1.537
31		3.820	2.357	1.052	.004	— .164	— .504
32			3.725	1.101	— .334	— .931	—1.115
33			3.820	2.054	.489	— .011	— .127
34		2.932	1.181	.395	— .127	— .982	—1.222
35		3.725	1.944	1.019	— .011	— .410	— .869
36			4.083	2.331	.851	— .322	— .575
37			4.275	2.811	.869	— .288	— .612
38			4.275	3.300	1.024	.945	— .015

XVII thus gives the location of every problem with reference to the median point of each grade distribution. The difficulty of any problem for any grade can be found by reference to this table.

4. Measuring the Distance Between the Grades

Thus far we have located each problem at the proper distance from the median point on the base-line of each grade distribution. We can now locate the difficulty of each problem for each particular grade. We also wish to know how difficult the problems are in general. We wish to know what will be the average position of each problem when placed on one linear scale. Before this can be done, we must determine the distances between the consecutive grade medians and we must establish a common zero point.

Three different methods have been used in this study to determine the interval between the grade medians. After the determinations derived from the three methods were satisfactorily weighted, the average was taken and used as the measure of the intergrade interval. For convenience these three methods will be called the "problem method," the "quartile method," and the "distribution method."

By the "problem method" the distance between the medians of two consecutive grades is determined by the difference in position each problem holds with reference to the medians of two consecutive grade distributions. For example, Table XVII shows that problem No. 2 is situated 1.788 P.E. below the median of the second grade and 2.384 P.E. below the median of the third grade. This makes a difference of .596 P.E. between the medians of the second and the third grades so far as this problem is concerned. Each problem will give a similar measure for the interval between any two consecutive grades. Table XVIII gives the P.E. intervals between the consecutive grades as determined from each addition problem.

It is interesting to note that as the problems increase in difficulty larger intervals tend to exist between the grade medians. This fact is most clearly brought out by Table XIX. In this table the determinations of the intergrade intervals from the various problems are divided into various groups. The group of determinations which is marked below −1.5 P.E. is the average of those determinations from Table XVIII which came from values lower than −1.5 P.E. in Table XVII; the group marked −1.5 P.E. to +1.5 P.E. is the average of those determinations obtained from values between −1.5 P.E. and +1.5 P.E.; the

TABLE XVIII

P.E. Intervals Shown Between Consecutive Grades by Each Addition Problem

PROBLEM NO.	INTERVAL II-III	INTERVAL III-IV	INTERVAL IV-V	INTERVAL V-VI	INTERVAL VI-VII	INTERVAL VII-VIII
1	—.254	.446	1.059	.678	—.188	.862
2	.596	1.011	.176	.367	.000	.145
3	1.408	.728	.055	—.055	.330	—.219
4	.412	.591	.944	.214	.335	.497
5	.713	.764	.951	.681	.358	.517
6	1.351	.677	.541	.737	.154	.358
7	1.594	.587	1.053	.737	.192	—.192
8	1.288	.808	.837	.512	.467	.875
9	2.236	.741	.454	—.044	—.170	.000
10	2.360	.768	.289	.496	—.250	.574
11	2.183	.814	.107	—.072	.092	.476
12	2.410	1.325	.654	.469	—.057	.342
13	1.754	.933	1.563	1.130	.000	.154
14	2.732	1.284	.834	.680	—.395	.000
15	2.606	1.286	.607	.437	.061	.391
16	2.920	.969	.848	.366	—.022	.348
17	2.395	.966	.412	—.058	.388	.110
18	2.562	1.159	.626	.126	.378	.508
19	2.738	1.196	.644	.428	.142	.602
20	2.686	1.626	1.254	.854	.503	.163
21	2.503	1.619	1.039	.941	.206	.009
22		1.467	1.155	.899	.365	.089
23		1.368	2.167	1.305	.366	.602
24			2.149	1.457	.373	.553
25		.862	2.227	1.211	.340	.655
26		.972	2.247	1.512	.324	.831
27			2.228	1.265	.643	.675
28			1.920	1.896	.440	.352
29			2.004	1.199	.648	.458
30		1.730	1.269	1.620	.463	.555
31		1.463	1.305	1.048	.168	.340
32			2.624	1.435	.597	.184
33			1.766	1.565	.500	.116
34		1.751	.788	.522	.855	.240
35		1.781	.925	1.030	.399	.459
36			1.752	1.480	1.173	.253
37			1.464	1.942	1.157	.324
38			.975	2.276	.079	.960

group marked above +1.5 P.E. is the average of the determinations obtained from values larger than +1.5 P.E. The group marked "select group" is the average of the determinations obtained from values between –2 P.E. and +2 P.E. The group marked "compositive" average is the average of the determinations in Table XVIII.

TABLE XIX

AVERAGES OF GROUPS OF DETERMINATIONS OF INTERGRADE INTERVALS AS MEASURED BY THE ADDITION PROBLEMS

GROUP	II-III INTER-VAL	III-IV INTER-VAL	IV-V INTER-VAL	V-VI INTER VAL	VI-VII -INTER-VAL	VII-VIII INTER-VAL
Below 1.5 P.E.	1.695	.842	.675	.508	.131	.347
—1.5 P.E. to +1.5 P.E.	.539	.959	.846	1.028	.509	.423
Above 1.5 P.E.	2.531	1.456	1.801	1.744	1.522[1]	.600[1]
Composite Average	1.866	1.099	1.155	.876	.300	.374
Select Group	1.429	1.093	.854	1.011	.456	.437

[1] These two values have been estimated.

Table XIX shows so far as this list of problems is concerned that the greatest difference between the medians of the different grades is brought about by the more difficult problems and that the least amount of difference is brought about by the least difficult problems. The one exception to this general statement is found in group --1.5 P.E. to +1.5 P.E. for the interval between the second and third year. The smallness of this determination is due to the small number of cases that happened to fall within those middle limits. These same facts are brought out in similar tables for the other fundamental processes. These results are in conformity with the results found by Dr. Trabue in his measurement with completion-test language scales.

Keeping in mind the fact that the greatest difference between the medians of any two consecutive grades is brought about by the more difficult problems and also that the least difference is brought about by the easier problems, it would seem that the best measure of the interval would be the average of those determinations that come near the median. It seems rather unfair that those problems at the extreme ends of a distribution should have equal weight with those near the middle of the distribution. In order to give more weight to the problems near the median of the distribution the average of those determinations in Table XVIII which were obtained from values –2 P.E. to +2 P.E. in Table XVII will be used in addition to the composite average in computing the final measure of the intervals between the grade medians. This last group of determinations is designated as the "select group" in Table XIX. Thus the composite average and the average of the determination of the "select group" are both measures of the intergrade intervals and both are derived from the "problem method" and will be used in the final determination of the intergrade intervals.

The second method of determining the distance between the grades was previously designated as the "quartile" method. If we have a normal surface of distribution, as we have assumed, the quartile of any distribution should be equal to the P.E. of that distribution. Therefore, if we divide the quartile of a distribution into the crude score intervals, we will get the interval between the medians of the grades in terms of P.E. Since for each interval between the grades there are two quartile measures, the average of the two quartiles is used as a divisor of the crude score interval between the grades.

Table XX shows the intervals obtained by this process.

TABLE XX

DETERMINATION OF QUARTILE INTERVALS BETWEEN GRADES

	GRADE II	GRADE III	GRADE IV	GRADE V	GRADE VI	GRADE VII	GRADE VIII
Median .	6.819	14.509	18.321	23.073	29.774	32.446	33.987
Quartile.	2.712	2.996	2.247	2.837	3.910	3.099	2.118

Score Interval...	7.690	3.812	4.752	6.701	2.672	1.541
Average Interval...	2.854	2.621	2.542	3.374	3.505	2.609
Quartile Interval...	2.694	1.459	1.869	1.986	.762	.591

This table is made from data taken from Table XII. The score interval or crude score between the second and third grade medians is the difference between the median number of problems solved in the second and third grades. The average quartile is the average of the quartiles of the second and third grade distributions. The crude score (7.690) divided by the average quartile (2.854) gives the quartile interval (2.694), the distance between the second and third grade medians. By hypothesis this quartile interval is in terms of P.E.

Third measure of the interval between the median of the different grades was by means of the "distribution method." It is based upon the amount of overlapping of the consecutive grade distributions. Table XII shows that there are pupils in the second grade that excel the median achievement of the pupils in the third grade. On the other hand, there are pupils in the third grade that do not reach the median achievement of the second grade pupils. Between the median of the second grade distribution (6.819) and the median of the third grade (14.509) distribution lie 46.22 per cent of the 489 cases in the second grade distribution; between the same medians lie 42.16 per cent of the 615 cases in the third grade distributions. Since these percentages are deviations from the median or 50 per cent they can be turned into P.E. values by referring to Table XIV. Thus, as determined from the second grade distribution, the interval between the second and third grade medians is 2.643 P.E. As determined by the third grade distribution the interval between the third and second grade medians is 2.099 P.E. Thus we have two direct measures for this same interval. Similarly, direct measures can be made for the intervals between each of the other succeeding grades.

By the same reasoning, between the second grade median and the fourth grade median lie 49.79 per cent of the second grade distribution; between the third grade median and the fourth grade median lie 35.98 per cent of the third grade distribution. Turning the percentages into P.E. we find that the fourth grade median is 4.256 P.E. above the second grade median and 1.601 P.E. above the third grade median. If we take the distance between the third and fourth grade median from the distance between the second and fourth grade medians, we get an indirect measure of the distance between the second and third grade medians. Thus 4.256 P.E. – 1.601 P.E. = 2.655 P.E., the indirect measure of the interval between the second and third grades.

Table XXI shows the percentage with P.E. equivalent of each grade lying between the median and the medians of the neighboring grades.

Various other indirect measures can be obtained in a similar way. Indeed, if one wished he could get a still further remote indirect measure of this same second and third grade interval by taking the distance between the third and fifth grade median from the distance between the second and fifth grade medians. This latter determination would, in the writer's opinion, be influenced so much by the extreme ends of the distributions that it should be given no weight in the final determination of the grade intervals.

In the final determination of the grade intervals by the "distribution method" the direct measures were felt to be the best measures, and hence they were given double weight while the indirect measures were given but single weight.

TABLE XXI

PERCENTAGE WITH P.E. EQUIVALENT OF EACH GRADE LYING BETWEEN THE MEDIAN AND THE MEDIANS OF THE NEIGHBORING GRADES

	II	III	IV	V	VI	VII	VIII
II% P.E.	46.22 2.634	49.79 4.256					
III% P.E.	42.16 2.099		35.98 1.601	49.38 3.709			
IV% P.E.	48.89 3.390	36.34 1.625		41.26 2.012	49.83 4.373		
V% P.E.		49.34 3.676	40.57 1.950		43.17 2.207	48.33 3.157	
VI% P.E.			48.96 3.428	38.44 1.775		18.43 .711	29.35 1.214
VII% P.E.				47.16 2.825	19.45 .754		14.32 .544
VIII% P.E.					36.30 1.622	18.47 .713	

TABLE XXII

DETERMINATIONS OF THE INTERGRADE INTERVALS FROM OVERLAPPING OF DISTRIBUTIONS IN ADDITION

DETERMINATION	II-III	III-IV	IV-V	V-VI	VI-VII	VII-VIII
Lower Indirect....		1.291	2.051	1.478	1.050	.868
Lower Direct......	2.099	1.625	1.950	1.775	.754	.713
Lower Direct......	2.099	1.625	1.950	1.775	.754	.713
Upper Direct......	2.634	1.601	2.012	2.207	.711	.544
Upper Direct......	2.634	1.601	2.012	2.207	.711	.544
Upper Indirect....	2.655	1.697	2.166	2.446	.670	
Total.........	12.121	9.440	12.141	11.888	4.650	3.382
AVERAGE.........	2.424	1.573	2.023	1.981	.775	.676

Table XXII shows the interval between the grades as determined by the distribution method.

Thus we have the results of the intergrade intervals as determined by the three different methods. All of the determinations have about the same general characteristics. The writer felt that the " select group " of determination, which is based upon the individual problems not varying over 2 P.E. in difficulty from the median problem of the grade, was the best measure. He also felt that the second best measure was the composite average based upon the difficulty of all the individual problems. Both the " selected group " determinations and the composite averages take into consideration the fact that the problems are not of the same difficulty. Hence it appeared to the writer that it might be fair to give these determinations increased weight in making the final determination of the intergrade intervals. Thus the " selected group " determinations were given a weight of three, and the " composite average " determinations a weight of two while the determinations from the quartile method and those from the " distribution method " were given but a single weight. Table XXIII gives the average of the determinations which is used as the final measure of the intergrade intervals in the construction of the Addition Scales.[1]

TABLE XXIII

FINAL DETERMINATION OF INTERVALS BETWEEN SUCCESSIVE GRADES

METHOD	II-III	III-IV	IV-V	V-VI	VI-VII	VII-VIII
Prob. select group.	1.429	1.093	.854	1.011	.456	.437
" " "	1.429	1.093	.854	1.011	.456	.437
" " "	1.429	1.093	.854	1.011	.456	.437
" composite av.	1.866	1.099	1.155	.876	.300	.374
" " "	1.866	1.099	1.155	.876	.300	.374
Distribution.......	2.424	1.573	2.023	1.981	.775	.676
Quartile..........	2.694	1.459	1.869	1.986	.762	.591
AVERAGE.........	1.877	1.216	1.252	1.250	.501	.475

[1] The other intergrade intervals as used in the final development of the other scales are as follows:

	II-III	III-IV	IV-V	V-VI	VI-VII	VII-VIII
Subtraction..........	1.646	1.073	1.392	1.233	.751	.820
Multiplication........		1.748	1.636	1.491	.822	.536
Division.............		1.014	1.554	1.368	.667	.659

5. Location of the Zero Point

Having determined the distance between the various grades, it is easy to locate all of the problems in terms of any grade median. However, if we wish to know the exact relation of one problem to another, if we wish to know how many times more difficult one problem is than another, it is necessary to find the location of each problem with reference to a common zero point.

Professor Thorndike in his *Mental and Social Measurements* (p. 16) in speaking of the definition of a zero point says, " The zero point may be *absolute,* measuring just not any of the thing, or *arbitrary,* meaning a point called zero though actually designating some amount of the thing. Thus the thing being temperature, 20° C is 20° above the arbitrary zero—the melting of ice—and 293° above the supposed absolute zero, just not any molecular motion in a gas."

The zero point in connection with any one of these scales is an arbitrary one. It means simply " the inability to solve correctly a single problem " as presented under the standard conditions for giving these tests. It does not mean that a child has absolutely no ability in addition or subtraction or multiplication or division. It is probable that if the problems had been presented orally to the pupils in the lower grades they would have solved more problems correctly. Some of the pupils who showed " inability to solve correctly a single problem " as the test was presented to them, no doubt would have solved a few problems if presented orally. Moreover, it should be added that zero ability in division (i.e., inability to solve a single problem) does not mean that a child will have zero ability in addition. It should also be added that we cannot say, as the scales are in their present condition, that a problem with a value of 1 in one process is equal in difficulty to a problem with the same value in another process. Each scale has its own zero point. We therefore cannot treat values as equal which have been developed from different zero points.

The method for locating the zero point is the same for all the fundamental processes, but I shall deal in detail only with addition. Table XII shows that of the 489 pupils in the second grade 44.88 per cent lie between those children who could not get a single problem and the median achievement for that grade

(6.819). Since this represents deviation from the median or 50 percentile, reference to Table XIV shows that a deviation of 44.88 per cent from the median of a normal distribution represents a distance of 2.322 P.E. This means then that the median of the second grade in addition is 2.422 P.E. above no score at all.

Table XII also shows that of 615 pupils in the third grade 49.34 per cent lie between those children who could not get a single problem and the median achievement for that grade (14.509). This would locate the median for the third grade 3.676 P.E. above the zero point. We have already determined that the median of the third grade is 1.877 P.E. (Table XXIII) above that of the second grade. By subtracting the distance the third grade median is above the second from the distance the third grade median is above the zero point, we get a measure of the distance the second grade median is above the zero point. Thus 3.676 P.E. minus 1.877 P.E. = 1.799 P.E., another measure of the distance that the second median is above the zero point.

Table XII also shows that the median achievement of the second grade distribution is 6.819 problems and that the quartile is 2.712 problems. Since we have assumed a normal surface of frequency, the quartile is equal to the P.E.; thus by dividing the median achievement by the quartile of the second grade we find that the median of the second grade is 2.514 P.E. above zero.

Similarly the median achievement for the third grade is 14.509 problems and the quartile 2.996 problems. By dividing the median achievement by the quartile we find that the third grade median is 4.843 P.E. above zero. Subtracting the distance the third grade median is above the second (1.877 P.E.) from the distance the third grade is above zero (4.843 P.E.) gives us the distance the second grade median is above zero (2.966 P.E.)

Thus we have four determinations of our zero point as follows:

From the second grade distribution	2.422
" " third " "	1.799
" " second " achievement	2.514
" " third " "	2.966
Average	2.425

The average of the four determinations probably represents a better measure of the distance than any single measure, so for the addition scale we shall use the point 2.425 P.E. below the second grade median as the arbitrary zero point.

Since we have determined the distance between the medians of the various grades (Table XXIII) and know that the second grade median is 2.425 P.E. above zero it is easy to determine the distance each grade median is above zero.

TABLE XXIV

DISTANCE THE MEDIAN OF EACH GRADE IN ADDITION IS ABOVE ZERO

1 = P.E.

GRADE	ABOVE ZERO	BELOW NEXT GRADE
II	2.425	1.877
III	4.302	1.216
IV	5.518	1.252
V	6.770	1.250
VI	8.020	.501
VII	8.521	.475
VIII	8.996	

Figure 18 (page 65) represents graphically the relations of the grade distribution to each other, the relations of the grade medians to each other and to the zero point, as determined by the values of Table XXIV and based upon the assumption that achievement in the solution of problems is distributed normally.

6. Referring All the Problems in Addition to Zero

Table XVII gives the value of each problem in addition for each grade. It shows, for instance, that problem No. 1 has a negative value of 1.217 P.E. in the second grade and a negative value of .963 P.E. in the third grade, etc. By reference to Table XXIV one finds that the second grade median is 2.425 P.E. and the third grade median is 4.302 P.E. above zero. By subtracting 1.217 from 2.425 and .963 from 4.302 we find that problem No. 1, in the second and third grades, is respectively 1.208 and 3.339 P.E. above zero. Wherever the value in Table XVII is positive instead of negative, as these just cited have been, add this value to the value which that particular grade median is above zero. For example, to determine the distance problem No. 8 is above zero in the second grade we add .187 P.E. to 2.425 P.E. Thus we find that problem No. 8 is 2.612 P.E. above zero in the second grade. Table XXV shows the location above zero of each problem in addition.

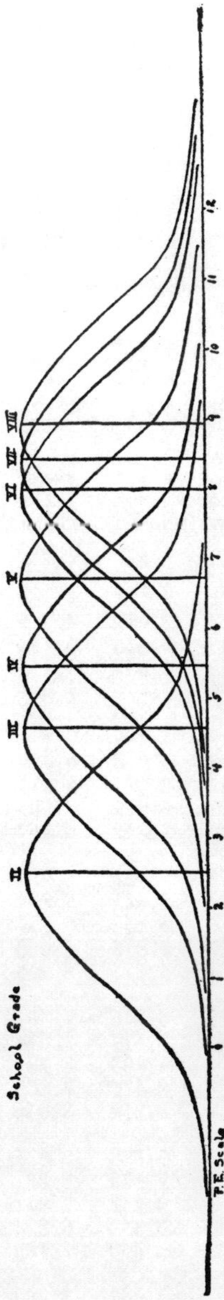

FIG. 18. RELATIONS OF GRADE DISTRIBUTIONS TO EACH OTHER IN ADDITION.

Since we have determined the location above zero of each problem in each grade we are now ready to determine the general value of each problem and to locate it on a linear scale. In order to do this we must find the position which best represents the difficulty of each problem.

TABLE XXV

LOCATION ABOVE ZERO OF EACH ADDITION PROBLEM

PROBLEM NO.	GRADE II	GRADE III	GRADE IV	GRADE V	GRADE VI	GRADE VII	GRADE VIII
1	1.208	3.339	4.109	4.302	4.874	5.563	5.176
2	.637	1.918	2.123	3.199	4.082	4.583	4.913
3	1.166	1.635	2.123	3.302	4.625	4.796	5.490
4	1.785	3.250	3.875	4.173	5.209	5.375	5.353
5	1.809	2.973	3.425	3.726	4.295	4.438	4.396
6	2.160	2.686	3.225	3.936	4.449	4.796	4.913
7	2.313	2.596	3.225	3.424	3.937	4.246	4.913
8	2.612	3.201	3.609	4.024	4.762	4.796	4.396
9	3.045	2.686	3.161	3.959	5.253	5.924	6.399
10	3.156	2.673	3.121	4.084	4.838	5.589	5.490
11	2.843	2.537	2.939	4.084	5.406	5.815	5.814
12	4.268	3.735	3.626	4.224	5.005	5.563	5.696
13	4.659	4.782	5.065	4.754	4.874	5.375	5.696
14	4.549	3.694	3.626	4.044	4.514	5.410	5.885
15	4.908	4.179	4.109	4.754	5.567	6.007	6.091
16	5.771	4.728	4.975	5.379	6.263	6.786	6.913
17	4.939	4.421	4.671	5.511	6.819	6.932	7.297
18	5.607	4.922	4.979	5.605	6.729	6.852	6.819
19	6.700	5.839	5.859	6.467	7.289	7.648	7.521
20	6.700	5.891	5.481	5.479	5.875	5.873	6.185
21	6.700	6.074	5.671	5.884	6.193	6.486	6.952
22		6.385	6.134	6.231	6.582	6.718	7.104
23		8.027	7.875	6.960	6.905	7.040	6.913
24			9.243	8.346	8.139	8.267	8.189
25		8.122	8.476	7.501	7.540	7.701	7.521
26		8.385	8.629	7.634	7.372	7.549	7.193
27			8.700	7.724	7.709	7.567	7.367
28			8.476	7.808	7.432	7.493	7.616
29			8.818	8.066	8.117	7.970	7.987
30		8.385	7.888	7.871	7.501	7.539	7.459
31		8.122	7.875	7.822	8.024	8.357	8.492
32			9.243	7.871	7.686	7.590	7.881
33			9.338	8.824	8.509	8.510	8.869
34		7.234	6.699	7.165	7.893	7.539	7.774
35		8.027	7.462	7.789	8.009	8.111	8.127
36			9.601	9.101	8.871	8.199	8.421
37			9.793	9.581	8.889	8.233	8.384
38			9.793	10.070	9.044	9.166	8.981

In making this determination it was felt that the truest value of any problem came from the distribution where the median achievement was nearest the location of that problem. It was also felt that those values which came from those distributions where the median achievements were farthest from the location of that problem should have little or no weight in the determination of the final value of the problem.

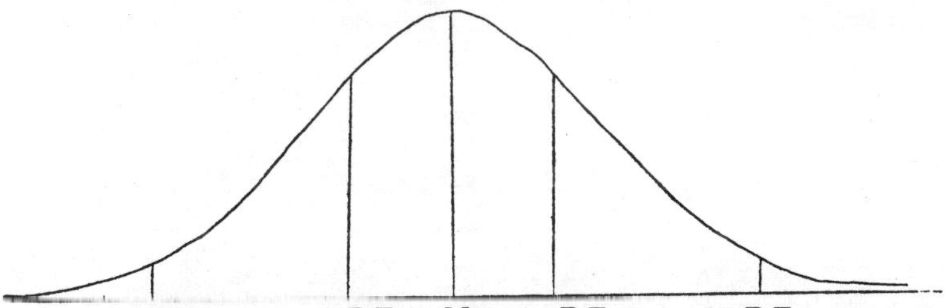

FIG. 19. SHOWING THE INFLUENCE OF VARIOUS PARTS OF A DISTRIBUTION IN DETERMINING THE GENERAL VALUE OF A PROBLEM.

Thus, as represented by Fig. 19, in computing the general final value of a problem in these various distributions, that value from each grade distribution is given double weight if the problem is less than 1 P.E. distance from the median achievement of that distribution; single weight, if it is more than 1 P.E. but less than 3 P.E. distance from the median achievement of the distribution; and not considered at all if the problem is more than 3 P.E. distance from the median achievement of the distribution. The average of all of these determinations is taken as the general difficulty of a problem.

Therefore, in making this final determination of the general value of any problem reference must be made to both Table XVII and Table XXV. Table XVII is used to determine the weight that shall be given to the value in Table XXV. For example, Table XVII shows that problem No. 1 is located at .963 P.E. from the median achievement of the third grade. It also shows that this same problem is located at more than 1 P.E. but less than 3 P.E. distance from the median achievement in grades two, four, and five. With these facts in mind weight the value given in Table XXV accordingly. The value in the third grade would have double weight, while the values in the second, fourth, and fifth grades would have single weight. The other values are disregarded. The average of the above determinations represents the difficulty of the problem.

Table XXVI gives the general value, i.e., distance above the arbitrary zero, for each problem in addition. These values are final and are used in locating the problems on the linear scale.

TABLE XXVI

FINAL VALUE OF ADDITION PROBLEMS

RANK	NO. OF PROBLEM	VALUE
1	2	1.23
2	3	1.40
3	5	2.50
4	7	2.61
5	6	2.83
6	8	3.21
7	1	3.26
8	4	3.35
9	10	3.63
10	11	3.78
11	14	3.92
12	9	4.18
13	12	4.19
14	13	4.85
15	15	4.97
16	17	5.52
17	16	5.59
18	18	5.73
19	20	5.75
20	21	6.10
21	22	6.44
22	19	6.79
23	23	7.11
24	34	7.43
25	26	7.47
26	30	7.61
27	27	7.62
28	25	7.67
29	28	7.71
30	32	7.71
31	35	7.97
32	29	8.04
33	31	8.18
34	24	8.22
35	36	8.58
36	37	8.67
37	33	8.67
38	38	9.19

NOTE.—These values are listed in Part I of this monograph. The graphic representation of the scale, i. e., the problem placed on a linear projection according to these final values, is also given in Part I.

TABLES OF CRUDE DATA FROM WHICH SCALES WERE DEVELOPED

TABLE XXVII

Distribution According to the Number of Subtraction Problems Solved

	Grade II	Grade III	Grade IV	Grade V	Grade VI	Grade VII	Grade VIII
35					6	14	34
34					8	42	45
33					8	51	56
32					23	76	68
31				2	27	68	80
30				1	25	79	75
29				0	41	86	57
28				2	39	90	48
27				13	38	66	33
26				13	46	76	15
25				19	68	70	18
24				33	70	34	7
23			5	57	63	46	9
22		2	14	61	59	52	0
21		1	13	84	36	23	5
20		7	42	101	36	27	2
19		24	51	72	17	6	2
18	1	26	54	68	24	5	1
17	0	24	54	51	14	4	
16	1	17	51	36	7	3	
15	1	28	58	17	0	0	
14	2	36	49	13	2	0	
13	8	52	60	7	1	0	
12	7	49	38	10	1	0	
11	11	54	27	8	1	1	
10	15	53	20	3	1		
9	23	58	14	2	0		
8	28	29	6	1	0		
7	33	26	11	1	0		
6	31	21	3	2	0		
5	34	13	5	0	0		
4	19	21	2	0	1		
3	13	10	1	2			
2	20	11	4	1			
1	30	21	6	1			
0	104	33	18	3			
No. Tested	381	616	606	684	662	919	555
Median	5.132	11.223	15.672	20.436	24.971	28.517	31.687
25 Per Cent.	.916	6.923	12.908	18.191	22.415	25.411	28.974
75 Per Cent.	8.063	14.306	18.509	22.492	28.295	31.313	32.945
Quartile	3.574	3.692	2.801	2.651	2.940	2.951	1.98

TABLE XXVIII

NUMBER IN EACH GRADE THAT SOLVED EACH PROBLEM IN SUBTRACTION CORRECTLY

PROBLEM NO.	GRADE II	GRADE III	GRADE IV	GRADE V	GRADE VI	GRADE VII	GRADE VIII
1	225	464	556	664	658	917	555
2	133	326	449	625	646	906	553
3	222	529	573	675	662	919	555
4	199	496	568	665	654	904	555
5	169	469	551	668	660	917	555
6	184	511	534	671	653	917	555
7	158	468	547	658	651	913	552
8	80	358	445	614	634	908	551
9	113	425	510	646	650	909	553
10	72	391	496	627	639	894	545
11	52	339	466	635	645	907	553
12	53	363	498	649	651	910	552
13	54	364	509	655	655	910	553
14	13	222	362	592	625	885	546
15	9	193	358	588	638	907	551
16	2	141	315	547	570	870	534
17	2	145	334	566	573	840	512
18	0	78	244	498	586	849	542
19	1	61	161	389	496	776	515
20		31	121	326	403	662	441
21		8	75	326	492	772	519
22		14	53	285	434	758	518
23		0	2	27	176	342	223
24		0	1	81	240	417	277
25		0	0	53	197	340	223
26		3	75	228	494	758	518
27			12	204	467	743	494
28			0	125	304	528	420
29			0	24	199	565	472
30			2	17	181	438	438
31				15	140	361	312
32				10	160	462	411
33				17	177	445	359
34				17	199	450	375
35				8	115	336	389
No. Tested	381	616	606	684	662	919	555

TABLE XXIX

Location Above Zero of Each Subtraction Problem

PROBLEM NO.	GRADE II	GRADE III	GRADE IV	GRADE V	GRADE VI	GRADE VII	GRADE VIII	FINAL VALUE
1	.782	1.748	1.781	2.416	2.735	2.936	3.431	1.501
2	1.691	2.650	2.877	3.201	3.528	3.953	4.093	2.645
3	.805	1.167	1.451	1.927	1.860	2.611	3.431	1.057
4	1.034	1.487	1.566	2.393	3.114	4.029	3.431	1.502
5	1.325	1.710	1.856	2.269	2.377	2.936	3.431	1.697
6	1.179	1.347	2.086	2.150	3.202	2.936	3.431	1.447
7	1.434	1.355	1.909	2.596	3.314	3.486	4.211	1.745
8	2.312	2.459	2.908	3.343	3.898	3.865	4.388	2.898
9	1.906	2.027	2.348	2.857	3.349	3.761	4.093	2.178
10	2.423	2.250	2.489	3.173	3.774	4.354	4.920	2.959
11	2.738	2.572	2.744	3.061	3.579	3.911	4.093	2.877
12	2.725	2.425	2.467	2.802	3.314	3.705	4.211	2.568
13	2.705	2.421	2.354	2.665	3.010	3.705	4.093	2.513
14	3.822	3.289	3.471	3.584	4.103	4.563	4.849	3.699
15	4.074	3.485	3.494	3.625	3.793	3.911	4.388	3.635
16	4.936	3.863	3.757	3.984	4.851	4.814	5.400	4.346
17	4.936	3.833	3.645	3.824	4.817	5.185	5.917	4.409
18		4.454	4.199	4.327	4.680	5.087	5.073	4.418
19	5.391	4.671	4.762	4.969	5.464	5.712	5.865	5.182
20		5.201	5.088	5.312	6.050	6.342	6.172	5.763
21		6.062	5.548	5.312	5.492	5.730	5.786	5.524
22		5.748	5.851	5.538	5.864	5.825	5.809	5.754
23			7.918	7.841	7.432	7.695	8.399	7.841
24			8.435	6.984	6.979	7.383	8.068	7.406
25				7.341	7.246	7.707	8.399	7.720
26		6.700	5.548	5.867	5.478	5.825	5.809	5.696
27			7.341	6.013	5.661	5.915	6.204	5.911
28				6.567	6.609	6.931	6.998	6.774
29				7.913	7.233	6.777	6.488	7.070
30			7.918	8.159	7.355	7.296	6.840	7.383
31				8.242	7.646	7.614	7.800	7.694
32				8.485	7.498	7.200	7.073	7.208
33				8.159	7.382	7.270	7.472	7.486
34				8.159	7.233	7.248	7.354	7.412
35				8.622	7.851	7.719	7.249	7.517

TABLE XXX

DISTRIBUTION ACCORDING TO THE NUMBER OF MULTIPLICATION PROBLEMS SOLVED

	GRADE III	GRADE IV	GRADE V	GRADE VI	GRADE VII	GRADE VIII
39					7	9
38					9	19
37				5	25	41
36				10	34	36
35				8	45	51
34				14	72	56
33				27	63	68
32				24	97	41
31				36	84	56
30			1	41	86	44
29			3	45	64	45
28			7	41	66	29
27			10	43	63	24
26			10	45	52	22
25		1	18	42	38	7
24		2	19	42	29	4
23		1	34	45	27	4
22		4	32	44	19	1
21		3	40	29	18	3
20		8	55	27	10	2
19		16	73	14	7	0
18	1	14	62	16	10	1
17	0	24	79	14	5	1
16	2	25	44	16	3	0
15	0	38	37	8	6	1
14	0	38	44	11	1	
13	7	49	27	5	1	
12	12	41	20	3	2	
11	17	42	21	3		
10	20	46	15	1		
9	17	45	5	1		
8	27	37	8	2		
7	27	29	4	0		
6	28	36	3	0		
5	44	36	8	2		
4	91	29	2	0		
3	52	19	0	0		
2	34	4	1	0		
1	22	4	0	1		
0	55	13	7			
No. TESTED...	456	604	689	665	943	565
MEDIAN.......	4.714	11.095	18.315	26.144	30.587	32.940
25 PER CENT ..	3.058	7.345	15.196	22.301	27.123	29.940
75 PER CENT ..	7.593	14.605	21.044	29.973	33.306	35.289
QUARTILE......	2.267	3.630	2.924	3.836	3.091	2.674

TABLE XXXI

NUMBER IN EACH GRADE THAT SOLVED EACH PROBLEM IN MULTIPLICATION CORRECTLY

PROBLEM NO.	GRADE III	GRADE IV	GRADE V	GRADE VI	GRADE VII	GRADE VIII
1	372	577	673	660	936	563
2	373	570	671	657	938	561
3	338	545	676	658	936	565
4	281	529	661	652	936	563
5	157	445	644	656	935	559
6	137	421	644	645	937	562
7	139	462	638	648	932	562
8	134	447	651	649	933	562
9	90	349	602	613	898	547
10	82	357	593	615	899	547
11	34	284	571	603	893	548
12	40	273	553	589	868	527
13	18	198	453	526	820	514
14	6	173	433	485	768	506
15	5	149	439	571	880	545
16	8	192	495	555	823	491
17	0	14	228	430	844	541
18	3	100	408	503	770	501
19	2	85	400	531	801	509
20	3	68	251	330	593	402
21	0	9	172	398	820	528
22	0	5	123	418	821	542
23	4	56	248	515	834	547
24	0	36	195	476	834	539
25	0	6	110	442	829	534
26	0	5	100	400	653	454
27	0	0	45	304	541	380
28	0	5	51	348	665	447
29	0	11	93	296	545	326
30	0	1	37	253	535	388
31	0	6	39	296	597	419
32	0	1	32	210	460	363
33	2	18	76	227	452	349
34		0	40	179	373	323
35		0	5	104	271	264
36		3	6	141	377	307
37		1	15	139	406	319
38		0	2	43	212	182
39		0	1	39	186	176
NO. TESTED...	456	604	689	665	943	565

TABLE XXXII

Location Above Zero of Each Multiplication Problem

PROBLEM NO.	GRADE III	GRADE IV	GRADE V	GRADE VI	GRADE VII	GRADE VIII	FINAL VALUE
1	.256	.825	2.017	2.823	3.645	3.886	1.050
2	.245	.982	2.094	3.120	3.468	4.181	1.107
3	.633	1.421	1.898	3.016	3.645	3.224	.872
4	1.154	1.626	2.396	3.389	3.645	3.886	1.582
5	2.187	2.399	2.730	3.166	3.717	4.374	2.380
6	2.364	2.574	2.730	3.677	3.563	4.004	2.713
7	2.347	2.268	2.830	3.585	3.942	4.004	2.675
8	2.394	2.385	2.605	3.534	3.838	4.004	2.616
9	2.855	3.047	3.276	4.363	4.820	5.078	3.783
10	2.948	2.998	3.366	4.332	4.805	5.078	3.789
11	3.725	3.451	3.566	4.504	4.891	5.035	4.089
12	3.598	3.518	3.711	4.678	5.205	5.602	4.261
13	4.188	3.999	4.372	5.265	5.618	5.836	4.706
14	4.891	4.177	4.486	5.562	5.964	5.957	5.046
15	4.986	4.353	4.456	4.871	5.066	5.138	4.723
16	4.737	4.041	4.120	5.022	5.596	6.154	4.727
17		6.297	5.623	5.907	5.429	5.262	5.721
18	5.234	4.777	4.630	5.438	5.948	6.029	5.242
19	5.529	4.934	4.672	5.223	5.751	5.915	5.194
20	5.234	5.134	5.491	6.481	6.799	6.995	6.296
21		6.558	5.975	6.094	5.618	5.579	5.889
22		6.910	6.338	5.977	5.611	5.227	5.826
23	5.097	5.301	5.506	5.346	5.516	5.078	5.375
24		5.644	5.826	5.619	5.516	5.326	5.625
25		6.789	6.450	5.834	5.553	5.454	5.825
26		6.910	6.544	6.083	6.540	6.555	6.290
27			7.220	6.626	7.011	7.159	6.973
28		6.910	7.120	6.381	6.489	6.623	6.580
29		6.450	6.611	6.671	6.996	7.536	7.002
30		7.614	7.359	6.915	7.038	7.101	7.066
31		6.789	7.319	6.671	6.784	6.861	6.850
32		7.614	7.458	7.176	7.332	7.281	7.290
33	5.529	6.128	6.786	7.074	7.366	7.379	7.069
34			7.306	7.379	7.679	7.555	7.504
35			8.618	7.965	8.122	7.947	8.020
36		7.159	8.481	7.652	7.668	7.660	7.656
37		7.614	7.961	7.667	7.546	7.581	7.647
38			9.250	8.711	8.408	8.509	8.533
39			9.575	8.784	8.552	8.551	8.609

TABLE XXXIII

DISTRIBUTION ACCORDING TO THE NUMBER OF DIVISION PROBLEMS SOLVED

	GRADE III	GRADE IV	GRADE V	GRADE VI	GRADE VII	GRADE VIII
36				2	6	5
35				0	9	21
34				7	25	33
33				10	38	47
32				11	48	59
31			1	18	74	51
30			1	29	67	62
29			0	25	86	63
28			3	42	74	41
27			3	46	77	51
26			2	40	73	30
25			7	45	58	29
24			13	51	58	11
23		1	20	41	60	11
22		1	27	42	38	10
21		1	41	47	28	6
20		3	49	39	29	7
19		2	53	40	26	1
18		4	47	30	8	1
17		4	50	26	19	0
16		14	48	18	14	1
15	1	10	43	16	8	2
14	5	19	77	13	6	
13	14	48	48	9	3	
12	7	47	46	11	2	
11	4	74	33	4	1	
10	12	66	21	2	2	
9	12	67	20	1	1	
8	13	58	11	0	1	
7	15	44	3	2	0	
6	21	37	2	0	0	
5	27	25	4	1	0	
4	12	25	4	2	0	
3	12	17	1		0	
2	10	12	3		1	
1	21	8	3			
0	32	18	1			
No. Tested...	218	605	685	670	940	542
Median.......	5.815	9.873	16.469	23.781	27.442	30.113
25 Per Cent..	2.125	7.211	13.401	19.813	23.800	27.519
75 Per Cent..	9.042	12.585	19.919	27.489	30.478	32.500
Quartile......	3.459	2.687	3.259	3.838	3.339	2.491

TABLE XXXIV

NUMBER IN EACH GRADE THAT SOLVED EACH PROBLEM IN DIVISION CORRECTLY

PROBLEM NO.	GRADE III	GRADE IV	GRADE V	GRADE VI	GRADE VII	GRADE VIII
1	91	302	506	547	822	470
2	110	325	544	623	912	533
3	98	369	548	578	865	526
4	100	365	571	656	929	541
5	87	451	657	662	934	542
6	138	500	656	661	932	541
7	124	485	656	644	913	532
8	110	428	649	653	930	538
9	76	415	649	655	931	535
10	24	44	315	493	859	527
11	42	271	473	612	917	536
12	96	446	607	582	824	485
13	93	453	595	605	856	511
14	12	144	462	618	893	534
15	4	42	254	496	791	506
16	0	39	415	512	818	502
17	4	31	180	458	779	491
18	4	178	463	578	857	521
19	0	78	313	379	670	433
20	0	37	283	402	621	401
21	0	16	229	433	731	476
22	0	40	267	450	725	485
23	5	25	132	433	675	451
24		0	91	260	640	438
25		8	63	373	638	420
26		1	143	352	658	426
27		6	155	401	678	464
28			18	208	364	323
29			18	221	442	320
30			6	209	491	397
31			18	207	420	313
32			10	179	379	299
33			0	74	265	313
34			5	115	321	254
35			1	30	158	136
36			0	28	123	139
No. Tested...	218	605	685	670	940	542

TABLE XXXV

LOCATION ABOVE ZERO OF EACH DIVISION PROBLEM

PROBLEM NO.	GRADE III	GRADE IV	GRADE V	GRADE VI	GRADE VII	GRADE VIII	FINAL VALUE
1	2.230	2.937	3.538	4.515	4.823	5.532	3.586
2	1.900	1.795	3.270	3.550	3.733	4.035	2.563
3	2.106	2.519	3.244	4.233	4.439	4.392	3.194
4	2.072	2.546	3.049	2.840	3.176	2.906	2.457
5	2.299	1.956	1.908	2.509	2.797	2.581	2.083
6	1.415	1.542	1.925	2.555	3.016	2.906	1.574
7	1.661	1.674	1.925	3.241	3.711	4.104	2.312
8	1.900	2.126	2.090	2.950	3.072	3.538	2.182
9	2.494	2.214	2.090	2.869	3.016	3.881	2.395
10	3.730	5.088	4.636	4.919	4.496	4.347	4.596
11	3.205	3.127	3.752	3.839	3.590	3.786	3.484
12	2.143	1.993	2.699	4.192	4.802	5.322	3.160
13	2.192	1.937	2.824	3.920	4.525	4.837	3.045
14	4.289	3.990	3.818	3.752	4.083	3.962	3.958
15	5.030	5.132	4.976	4.901	5.041	4.947	4.982
16		5.178	4.088	4.788	4.852	5.036	4.671
17	5.030	5.358	5.427	5.145	5.113	5.228	5.263
18	5.030	3.736	3.810	4.233	4.515	4.567	4.058
19		4.610	4.647	5.609	5.688	5.938	5.304
20		5.226	4.813	5.483	5.906	6.227	5.564
21		5.814	5.123	5.300	5.387	5.453	5.357
22		5.167	4.901	5.195	5.421	5.322	5.157
23	4.877	5.512	5.773	5.300	5.667	5.754	5.481
24			6.136	6.277	5.824	5.890	6.038
25		6.233	6.458	5.642	5.833	6.061	5.911
26		7.533	5.688	5.762	5.740	6.005	5.782
27		6.383	5.602	5.483	5.653	5.605	5.579
28			7.368	6.586	6.948	6.821	6.868
29			7.368	6.507	6.634	6.844	6.762
30			7.993	6.582	6.440	6.263	6.428
31			7.368	6.595	6.720	6.893	6.826
32			7.745	6.777	6.882	6.987	6.882
33				7.666	7.377	6.893	7.241
34			8.130	7.258	7.130	7.296	7.222
35				8.369	7.949	8.177	8.168
36				8.417	8.185	8.153	8.227